O9-AHW-194

Diversity and the Tropical Rain Forest

DIVERSITY AND THE TROPICAL RAIN FOREST

John Terborgh

SCIENTIFIC AMERICAN LIBRARY

A division of HPHLP
New York

Library of Congress Cataloging-in-Publication Data

Terborgh, John, 1936–
    Diversity and the tropical rain forest / John Terborgh.
       p. cm.
    Includes index.
    ISBN 0-7167-5030-9
     1. Rain forest ecology. 2. Biological diversity—Tropics.
   3. Rain forest conservation. 4. Biological diversity conservation—
Tropics. I. Title.
    QH541.5.R27T47 1992
   574.5′2642′0913—dc20   91-30053
                                                            CIP

ISSN 140-3213-5026-0

Copyright © 1992 by Scientific American Library

No part of this book may be reproduced by any mechanical, photographic, or
electronic process, or in the form of a phonographic recording, nor may it be stored
in a retrieval system, transmitted, or otherwise copied for public or private use,
without written permission from the publisher.

Printed in the United States of America

Scientific American Library
A Division of HPHLP
New York

Distributed by W. H. Freeman and Company
41 Madison Avenue, New York, New York 10010 and
20 Beaumont Street, Oxford OX1 2NQ, England

   3 4 5 6 7 8 9 0 KP 9 9 8 7 6 5 4

This book is number 38 of a series.

Contents

Preface

My senses snap to a state of full alert, straining to receive every signal from the night outside my tent. While still fully asleep, I had been aware of hearing an ominous crack, whose resonant undertones signal perhaps the greatest danger of living in the rain forest—being crushed by a falling tree. Having heard this sound many times before, I well knew that it heralds the first weakening of a venerable giant. Only a sheet of ripstop nylon protects me in the darkness. Where was the failing tree? How far? In which direction? Should I bolt out of the tent into the night in the hope that I will be less vulnerable as a moving target?

I don't have to weigh my options for long. A moment later another crack rings through the forest, followed by another, and then a whole series in accelerating staccato. The cracks quickly coalesce into a wrenching groan as the massive trunk begins to topple earthward in a crescendo of rushing leaves, snapping branches, and popping vine stems. Time slows for a few seconds as the rush of sound races to its thunderous conclusion and the great trunk meets its resting place on the forest floor. The ground shudders beneath me. I am now wide awake, heart pounding, eyes searching the darkness for any further sign of danger. Life in the rain forest begins unobtrusively, but often ends violently. I am hoping mine doesn't follow the same pattern.

What draws me to this distant retreat, so far beyond the fringes of what we call "civilization," and so alien to modern urban man? If I had to choose one word, it would be peace. Being daily witness to the endless cycle of life and death brings a reassuring sense of symmetry and continuity. The lives of the plants and animals that share the forest are inextricably linked in a web of interactions—a web held together by a system of checks and balances that we are only now beginning to understand. The orderly and predictable functioning of the forest provides a welcome counterbalance to the tumultuous, irrational, and often self-destructive affairs of our own species. The human world is racing along an exponential trajectory toward an uncertain future, and this I find profoundly disturbing. The rain forest offers certainty and sanity. Just as an astronomer finds beauty and mystery in the heavens, so, as a biologist, I find these rewards in the tropical forest. Admittedly, the rewards are personal ones, but they are rich and satisfying, so much so that I have returned to the

Amazon each year for twenty-eight years to savor peace, and to indulge my scientific curiosity.

I was first drawn to tropical ecology just after completing my Ph.D. in 1963, when a colleague and I made a celebratory expedition to South America. The subsequent years have been exciting ones scientifically, as an era of exploration and discovery has given way to an era of inquiry that examines how the tropical forest ecosystem functions and why it evolved to be the way it is. Unlike physics, in which discoveries of major principles are traced as far back as Newton or Galileo, the science of tropical ecology is largely new. It acknowledges a major debt to the eighteenth-, nineteenth-, and twentieth-century explorers who discovered and named the myriad organisms, but the formulation of general principles has come only recently. A number of these principles will be the subjects of this volume. They explain, for example, how closed nutrient cycles support highly productive forests on soils of extremely low fertility; why biological diversity is concentrated in the tropics; and why animals are indispensable to perpetuating the diversity of plant communities and vice versa.

No book can be everything to all readers. This one pays only passing homage to the myriad insects, spiders, and lesser invertebrates that populate the tropical forest. Trees and vertebrates give me more than enough to say, and invertebrates have never been my forte. So, with apologies to the entomologically inclined, the main staple shall be the larger things that have always held for me a special fascination.

Even as tropical biologists are caught up in the excitement of discovering basic principles, we work under a darkening cloud. The alarming destruction of tropical habitats suggests that the last opportunities to make basic discoveries about undisturbed ecosystems may occur within the next generation. The realization that external forces may impose a time limit on one's science engenders a feeling of panic akin to claustrophobia. Yet, thus far, it is almost a private panic. The public has become keenly aware of the threat of deforestation, but not of the foreclosing of scientific frontiers that deforestation implies. Least of all has a sense of urgency permeated the granting agencies, for tropical biology enjoys no particular priority in the intense competition for research funds. Instead, tropical biology languishes as a tiny and almost voiceless contestant in a throng of more powerful contenders pressing Congress to support the sequencing of the human genome, a manned space station, or a superconducting supercollider. But regardless of official inattention, the clock continues ticking. My fervent hope is that we wake up before it is too late.

Tropical deforestation is all the more disturbing because it threatens the biological diversity that is the basis of our existence. Of all the crimes and errors of our time, to paraphrase Professor Edward Wilson of Harvard, the one that will never be forgiven by future generations is the loss of our biological heritage. If current trends continue unchecked, the world's great tropical forests will be gone in thirty years, as impossible to bring back as the dinosaurs or the dodo.

May I suggest that any prospective readers who have as little time for enjoying books as I do begin with Chapters 8 and 9. These spell out the bottom line—the threats to tropical diversity and what steps might be taken to ameliorate them. My recommendation would then be to continue with Chapters 1 and 2 for a general overview of tropical forests and an account of the nearly closed nutrient cycles that make them so especially vulnerable to disturbance. If time permits and curiosity compels, then the scientific heart of the book will be found in Chapters 3 through 7, where I examine the many processes, both ecological and evolutionary, that contribute to tropical biodiversity.

It is my pleasure to thank a number of colleagues who read preliminary drafts of chapters and offered helpful comments, foremost among them Mercedes Foster, who read and criticized nearly all the chapters, but also Peter Ashton, Alwyn Gentry, Stephen Hubbell, and Carel van Schaik, each of whom read one or more of the chapters or helped by providing unpublished information. Michael Riley printed, photocopied, wrote letters, made telephone calls, and generally made himself indispensable in myriad ways. For whatever grace and flow may ease the text, I am indebted to Susan Moran, a skillful editor whose gimlet eye and unflagging professionalism were sometimes bruising to the ego, but nearly infallible in their precision and good taste. The remarkable photographic collection that embellishes these pages is a tribute to the taste and industry of Travis Amos. Travis proved himself a veritable wizard in his ability to conjure, literally from the far corners of planet, just the right photograph for each of dozens of erudite themes. For me, it has been a special privilege to go through this experience, not only with Susan and Travis, but with Joseph Ewing, Philip McCaffrey, and the rest of the staff at Freeman. I wish them all well in future endeavors.

John Terborgh
August 1991

Diversity and the Tropical Rain Forest

The Biological Exuberance of the Tropics

Before the publication of a two-page article in 1982, biologists, if asked to give an estimate of the number of species on earth, would have replied almost in unison with the number two million. That number would have been arrived at by adding half a million suspected undiscovered species to the 1.5 million or so species that had been formally described. The estimate of undiscovered species would represent the collective wisdom of experts in various groups of arthropods (insects, spiders and their allies, crustaceans, and some lesser groups). Because arthropods are by far the largest group of organisms, they include most of the species that are still unknown.

Rising mists waft gently over a cloud forest in Costa Rica.

That estimate was shattered after Dr. Terry Erwin, an entomologist from the Smithsonian Institution, began to investigate the arthropods that inhabit the tropical forest canopy. This last biological frontier had been almost totally neglected because of the danger and difficulty of ascending into trees that tower 40 to 50 meters above the ground. Erwin had developed a technique that allowed him to fumigate the crowns of individually selected trees. Overcome by a fog of biodegradable pyrethrin, insects, spiders, and other invertebrates came tumbling down and were collected on plastic sheets spread over the ground. Erwin discovered that there are a lot of arthropods in the tropical forest canopy, far more than anyone had imagined, perhaps as many as 30 million species.

Erwin's drastically revised estimate of the number of species on earth is based on the number of beetles collected after fumigating the crowns of 19 *Luehea seemannii* trees in a Panamanian forest. His reasoning runs, in brief, as follows. Over a three-season sampling program, the 19 trees yielded some 1200 species of beetles. Erwin then conservatively estimated at 13.5 percent the proportion of these 1200 beetles that were host-specific, that is, having at some stage of their life cycle an absolute requirement for *Luehea seemannii* trees. The remaining 86.5 percent of the collected beetles were then regarded as transients, or species with broad host requirements able to feed on other tree species as well. If each of the 70 species of canopy trees found in the average hectare of this forest were to have similar numbers of host-specific beetles (162), and if one conservatively assumes that all the transient and generalist species were discovered in the

Beetles of many sizes, shapes, and colors inhabit the crowns of tall trees in the rain forest canopy. Smithsonian entomologist Terry Erwin has recently discovered that this heretofore unexplored environment may harbor many millions of beetles and other arthropods unknown to science.

Luehea samples, the total number of beetle species to be found in the canopy of one hectare of Panamanian forest comes to about 12,500.

But beetles are only one group of arthropods, albeit the largest. They comprise about 40 percent of all described arthropod species. If one then multiplies the total number of beetle species by 2.5 to arrive at an estimate for all canopy-dwelling arthropods, and then increases that number by another third to account for the arthropods of the forest floor, the estimate for the total number of arthropod species in a hectare of Panamanian forest becomes more than 41,000!

That a single hectare of forest should harbor so many species is astounding. Nevertheless, one hectare offers but an infinitesimal sample of the world's diversity of life. To arrive at an estimate for the global total, we begin by noting that there are roughly 50,000 species of trees in tropical forests worldwide. If each one of them were to support 162 host-specific beetles, then there may be as many as 30 million species of tropical forest arthropods. With this one deductive leap, the estimate of the world's endowment of biological species has been revised upward by a factor of 15!

Without wishing to dispute Erwin's arithmetic, it is safe to say that such a wild extrapolation is bound to be in error. A common tree is more likely to support large numbers of host-specific arthropods than a rare one, and *Luehea seemannii* is a common tree in the Panamanian forest where Erwin worked. Moreover, it is a fairly large tree. If one considered only tropical forest tree species that were as large, common, and widespread as *Luehea seemannii*, the world total would be far less than 50,000. My own hunch, therefore, is that global species diversity will eventually prove to fall considerably short of 30 million species.

The Extraordinary Diversity of the Tropics

The discovery of thousands of hitherto unknown beetles in the tropical forest canopy should not have been a great surprise to biologists. Nature reaches its fullest expression in the tropical forest, whether measured by sheer numbers of species or by the complexity of their interactions. Nurtured for eons in a spacious and physically benign environment, tropical life has evolved an exuberant variety of species that has captured the fascination of scientists since the time of Darwin. The imposing diversity of trees, birds, insects, and other groups has led to the formation of intricate predator-prey webs and the refinement of *interspecific* interactions, interactions between two or more species that may include competition, mimicry, parasitism, and mutualism (the latter is an interdependency, such as that between the alga and fungus that form a lichen). In the prevalence of such interactions, tropical environments stand in contrast to the high latitudes, where smaller numbers of species are adapted, not so much to one another, as to the stresses of a long, dark, and frigid winter.

Forests comprise the natural vegetation over much of the region lying between the tropics of Cancer and Capricorn (23.5 degrees north and south of the equator, respectively). The intense evaporative power of the overhead sun draws moisture from the soil by way of the foliage of trees in a process called transpiration. Transpiration is nature's recycling mechanism for water. It charges the atmosphere with abundant moisture that is swept up in towering

Two actors in a complex web of interdependency: a passion vine butterfly *(Heliconius clysonymus)* visits a flower of a male *Psiguria* vine. Nectar from the vine supplies energy to the butterfly, while the butterfly acts as matchmaker, carrying pollen between male and female plants.

cloud banks before condensing into an afternoon downpour. The mild nonseasonal environment is congenial to the growth of trees and the myriad organisms that directly or indirectly depend on trees for food and shelter.

To the untrained eye, the fabled diversity of the tropical forest may be undiscernible. Aside from the buttressed trees, a few palms, lianas, and the occasional epiphyte, the forest looks the same in all directions. Although each plant may be a different species, a glance at the foliage does not reveal the differences. The trees are mostly small. Out of a hundred, one or two may be of impressive girth, but their crowns rise out of sight, hidden by several intervening layers of foliage. There is a strong sense of verticality, for the predominate view is of dozens of slender trunks disappearing into the leafy canopy above. One hears the chirping and droning of insects, and occasional bird calls, but the sources of the sounds seem invisible. At this point the traveler must wonder if tropical diversity is just fantasy.

The diversity is there, but most of it is hidden, often in subtle ways. The myriad trees with seemingly identical foliage can be distinguished, but one has to be versed in the fine points of plant anatomy. Hundreds of bird species can occur within a few hectares, but the numbers of each species are only a tenth of the norm in northern woods. Even an astute observer does not discover all the species in a day, a week, or even a year. Although I have been three or four months each year in the Amazon for 25 years, I have seen a jaguar five times, a puma only once, and I have never seen a harpy eagle or several other creatures of lesser fame.

The rewards of perseverance are almost boundless. The forest at my research site in Amazonian Peru contains 200 species of trees per hectare. A hundred hectares provides breeding habitat for 230 species of land birds, more than breed in most U.S. states. Ninety species of frogs and toads can be found in a few square kilometers, more than in all 50 states. When fumigated, the crown of a single large tree yielded 54 species of ants, more than have been recorded in the entire British Isles. A colleague has collected over 1200 species of butterflies at a locality nearby. Such extravagant diversity characterizes almost any group of organisms inhabiting the tropical forest.

Accumulating these species counts required many years. The specialist must learn the organisms, what habitat each occupies, when it breeds or flowers, and how to discriminate numerous related forms. In most cases, there are no field guides, and so identifications cannot be made on the spot. Lists of species are accumulated slowly as the product of repeated back-and-forth trips between the field site and big-city museums. The most comprehensive knowledge is in the heads of a few dedicated experts who can know well only a few sites at most. Tropical biology is truly an

Ocotea americana

Ocotea cooperi

Leaves of two trees in the laurel (avocado) family. Field botanists must become adept at distinguishing related species by discriminating subtle details from the ground through binoculars. The leaf on the left has fewer, more widely spaced veins and a longer petiole than the one on the right.

arcane art, and its practitioners are few relative to the size of their task. For this reason data are scant and tend to refer to a mere handful of well-studied sites.

The Scientific Investigation of Biodiversity

Two kinds of crows live in my backyard in Durham, North Carolina. They are both black, and to the eye are indistinguishable, but they make distinct sounds. From these sounds I can easily tell them apart, and whether by sound or by other means, they tell each other apart, for hybrids between them are unknown. By avoiding reproductive confusion, the American crow and the fish crow conform to the biological definition of distinct species.

Species are the units of evolution. They are populations of organisms that interbreed among themselves. Interbreeding ensures genetic coherence and an independent evolutionary destiny. Different species, by definition, are reproductively isolated from each other, although in some cases they may bear a close resemblance, as do the two species of crows.

Why are there just two species of crows in Durham? Some parts of Europe have several, while over most of North America there is only one. How many species of plants and animals inhabit the earth? Why do more of them live in the tropics than in the temperate zone? These are the kinds of questions that fascinate biologists trying to understand the complexities of *biodiversity*. This term, which was unknown to me a few years ago, is now a buzzword of the conservation movement. It refers generally to the variety of life, but can best be equated with the variety of species.

My goal in the ensuing chapters shall be to explore the biology of diversity as exhibited in its most exuberant manifestation—the tropical forest. The basic question is one a child might ask: Why are there so many species? Despite its disarming simplicity, this is a question no scientist can answer, and it is far too sweeping to be attacked directly.

If the empirical answer to the question "How many species are there?" is not known even to the nearest 10 million, then the theoretical side is even weaker. A comprehensive theory of diversity would provide a means for estimating the number of beetles or birds from first principles, but no such theory exists. Scientists have described approximately 9000 bird species, for example, and the declining rate of new discoveries worldwide suggests that this figure is within 1 percent of the actual total. Yet no one can even begin to offer reasons for why there are 9000 bird species and not 18,000 or 4500. By the same token, an equally profound mystery is why there can be as many as 300 species of trees in a single hectare of forest near Iquitos, Peru. On fundamental points such as this, our ignorance is almost unlimited.

The best that can be said is that the total number of species of any group is subject to regulation by a system of evolutionary checks and balances. We know this because we know that the number of species in many groups that are well represented in the fossil record has not systematically increased or decreased for millions of years. Over such long time spans individual species, even genera and families (hierarchical groupings of species), appear and disappear at more or less predictable rates. Moreover, the fossil record is punctuated by periodic crises, mass extinction events that drastically lower diversity. Yet, within 5 or 10 million years after these setbacks, diversities frequently return to approximately their former levels, although nearly all the species are different. Such evidence provides tantalizing hints that a master theory of diversity is not a vain hope, but only awaits a deeper understanding of the fundamental regulatory processes.

Greater success in understanding diversity has been achieved by asking more modest com-

A yellow-shouldered bat *(Sturnira lilium)* hovers as it plucks a *Solanum* fruit. New World fruit bats are atypical of fruit-eating animals in their habit of specializing on particular families of food plants. Members of a related genus, *Artibeus*, feed principally on figs, while those of a third genus, *Corollia*, feed on relatives of the black pepper plant.

parative questions; for example, Why do tropical forests in general contain more tree species than their temperate counterparts? Or, for that matter, why do they contain more birds, mammals, reptiles, or almost anything else? Here the biologist is challenged only to explain why one number should be larger than another, not to account for the absolute magnitude of either. Even this lesser task is by no means simple, because it requires one to distinguish between processes that accommodate diversity and processes that generate it.

A simple thought experiment should serve to make this critical distinction. Imagine two identical islands of diversified topography, each supporting 100 tree species. In principle, the species could be distributed in a large number of ways. For simplicity, assume that on one island the species are all specialists, one thriving on coastal sands, another on rocky mountaintops, a third in freshwater swamps, and so on. Suppose we were to sample the tree diversity of this island by counting the number of tree species in a randomly selected hectare. In this imaginary world of specialists, a hectare would be likely to include just one species, or at most a few if it happened to fall on a particularly complex piece of topography. Now imagine that the second island carried a community of generalists that could grow anywhere. In this equally improbable scenario, sample hectares would routinely pick up 80 or 90 species, and conceivably all 100. On the first island, the measured tree diversity would be close to zero; on the second, it would be at the highest possible level consistent with the probabilities of sampling.

What is the difference? Because the same number of species occurs on both islands, the difference lies in their biological properties, whether they are narrowly or broadly adapted to the available range of environmental conditions. It will also matter whether they compete strongly or weakly, as strong interspecific competition leads to reduced local diversity. This mental exercise suggests that such biological considerations can potentially regulate diversity.

To carry the thought experiment one step further toward reality, imagine now that we are no longer considering just a pair of islands, but instead an entire archipelago, each island identical except that it may carry from as few as 10 to as many as 1000 species of trees. If we were to compare sample hectares on several of these islands, we could no longer attribute differences in diversity solely to the biological properties of the species. Some of the variation could simply be a reflection of the number of species present. If the biological attributes of the species were constant over the whole archipelago, then the number of species present on each island would primarily determine the numbers counted in the sample hectares.

Here we have an entirely different situation, one that is less dominated by biological considerations. The number of species present has mostly to do with other factors. An island located near a mainland would be invaded by new species more often than a far island; one that had recently experienced a major catastrophe, such as a fire or a volcanic eruption, might have suffered extinctions. Ultimately, the species had to evolve at some point in space and time, and some regions of the archipelago could be more conducive than others to the origination of new species. These explanations relate to geography, history, and evolutionary processes, not to the biological properties of already established species.

This great dichotomy between the biological factors that influence the coexistence of species in diverse communities, and the evolutionary factors that regulate extinction and the origination of new species lies at the heart of scientific efforts to understand biological diversity. Although the two sets of factors, ecological and evolutionary, are distinct in principle, in practice it is difficult to distinguish their effects because they are interrelated in complex ways. Unraveling and exposing these interrelationships will be a major objective in the ensuing chapters.

Ecological processes operate within the lifetimes of individual organisms, hence usually on small spatial scales, while evolutionary processes operate over the grand sweep of geological time and take place on the scale of continents. These fundamental distinctions suggest that the organization of our inquiry into tropical diversity be based on contrasting scales of time and space. Our inquiry begins with an overview of the results of evolution, as expressed in large-scale phenomena—the patterning of tropical vegetation on climatic gradients, the adaptations of plants to a wide range of soil conditions, and the contrasting levels of diversity found in temperate and tropical forests. Smaller scales then become appropriate as I attempt to answer such questions as, How can as many as 300 tree species coexist in a single hectare, and how do more than 200 bird species find distinct ways to subsist in a square kilometer of forest? The focus then returns to evolution as I consider the mechanisms that generate diversity and the question of whether evolution has generated consistent forms of community organization on separate continents. Ecology reappears once more toward the end when I address the timely question of how to conserve diversity in a global environment that is under pressure to supply ever greater flows of goods and services to our own species. I conclude with some thoughts on whether tropical forests can be managed sustainably for human benefit without compromising their immemorial roles as repositories of biodiversity.

A field biologist is dwarfed in the buttress folds of a giant kapok tree *(Ceiba pentandra)* in Peru. Such trees contribute to the grandeur and mystique of virgin tropical forests, but most of the trees in these forests are comparatively slender.

Tropical Forests and Climate

The adventure-story stereotype of a tropical forest as an impenetrable thicket of vines and bizarre foliage, teeming with hazards to human life, is grossly exaggerated. Tropical forests come in many guises. Some are possessed of grandeur, others of beauty; a few may seem extravagantly exotic, while many are hardly more remarkable than the wildlands in some national parks of the temperate zone. A good many are rather unprepossessing. In fact, there are few fail-safe criteria that distinguish a tropical from a temperate forest.

Picture books emphasize the exotic: towering buttresses that dwarf a human in their fold, lianas (woody vines) as thick as one's leg, and festooning epiphytes (plants that grow atop other plants). It is true that these and some other conspicuous plant adaptations are characteristic of tropical forests. What is less well known is that buttresses, lianas, and epiphytes may all occur in temperate forests. Ancient beeches and oaks in one of the few remaining virgin forests of eastern North America can have surprising buttresses, for example. Grapevines in the rich bottomlands of my native Virginia can attain the girth and pendant quality of tropical lianas. Mistletoe, Spanish moss, and resurrection fern can crowd the branches of venerable oaks in the southeastern United States, creating a visual impression matched only in the most exotic tropical forests.

To the practiced eye, there are other features that better serve to signal a low-latitude forest. A more complex vertical organization is one. When the vegetation is viewed in cross section, as along a fresh road cut, as many as four or five superimposed crowns of different species may be counted from top to bottom, whereas two or three would be the maximum in a temperate forest. An even more subtle, but highly reliable indicator would be the lack of perennating structures—specialized organs, such as bulbs and protected buds, that allow many temperate plants to survive the cold and drought of winter. Despite the paucity of universally diagnostic features, most tropical forests, especially the wetter ones, possess a distinctive, uniquely tropical ambience.

Tropical forests take many forms, largely controlled by variation in rainfall, temperature, and seasonality, although soil conditions may also play a role. The climate of the great region between the tropics of Cancer and Capricorn is distinguished by a steady year-round temperature, but, apart from this one defining feature, a variety of climates are possible. Annual rainfall may vary from less than 10 millimeters along the Peruvian coast to more than 10 meters along the Colombian coast just a few hundred kilometers to the north. Mean annual temperatures range from nearly 30°C (86°F) in hotspots like Djibouti or Darwin, Australia, to below freezing atop the glaciated peaks of the Andes. As we shall see later, the annual cycle of seasonal change is also a highly significant

feature of tropical climates, but the seasons are characterized by variation in rainfall rather than temperature.

It is possible to generalize about the patterns of vegetation controlled by climate because of an evolutionary phenomenon known as convergence. The idea is really quite simple. If natural selection operates on the organisms living in a given environment for long enough (long enough being perhaps millions or tens of millions of years), the organisms will become highly adapted to the particular features of that environment. A prediction derived from this central tenet of evolutionary theory would be that unrelated organisms living in an identical environment somewhere else—say, on another continent—could be expected to show similar, if not identical, adaptations.

Surprisingly, given the naive simplicity of this prediction, compelling evidence of convergence is apparent in the simple gestalt (often termed physiognomy) of vegetation formations. Convergence presumably provides the underlying reason for why a rain forest in Cameroon can hardly be told from one in Brazil by anyone other than a trained botanist. We see further evidence in such global consistencies as the presence of spiny and succulent plants in deserts, the ubiquity of lianas, buttresses, and epiphytes in rain forests, and the possession of bulbs, corms, and tubers by herbaceous perennials of seasonal temperate environments.

Within the tropics, the timing of leafing, flowering, and fruiting in relation to the annual cycle, referred to as phenology, is closely related to the seasonality of rainfall. Evergreen forests occur where there is no dry season, or only a short one lasting at most a few months.

Translucent ferns and bromeliads embellish the branches of a tree in a Venezuelan cloud forest. Epiphytes such as these abound in perpetually moist mountain forests, but are relatively infrequent in most lowland forests.

Where dry and wet seasons are of approximately equal duration, deciduous forests are the norm. The convergence of phenological patterns around the tropics is so consistent that an experienced fieldworker can make penetrating deductions about the climate of a locality by examining the vegetation, or penetrating deductions about the vegetation from knowledge of the climate.

Tropical forests exist wherever the climate is congenial. From warm, humid lowlands along the equator, forests extend into drier, more seasonal lands north and south, and up mountain slopes to chilly, cloud-draped alpine zones. Declining moisture or temperature eventually sets the limits at which trees yield to lesser plants. Thus, even on the equator there are treeless zones, such as the summit of snow-clad Mount Chimborazo in Ecuador, or the scorching desert plains of southern Somalia.

Before the interventions of our own species, such places as these were only minor anomalies in an otherwise verdant carpet of trees that flourished almost everywhere the equator crosses land. Indeed, tropical forests once covered more than 20 percent of the earth's land surface. Now, about two-thirds of the original forest has already been replaced with other kinds of vegetation—croplands, pasture, less-complex secondary forests, tree plantations, and what can be called wasteland, because it serves no useful purpose for either man or nature.

Vegetation and Temperature

At most localities within the tropical zone, there is little variation in temperature on any time scale longer than a day. Annual means for most localities at low elevation fall in the range of 23 to 27°C. Indeed, the climatic tropics (as distinct from the geographical tropics) are often defined as those portions of the earth where the mean daily temperature excursion exceeds the range circumscribed by monthly means. Seasonal variation in temperature slowly increases away from the equator, although annual mean temperatures remain almost constant out to a latitude of nearly 25°, because warmer summers tend to offset cooler winters. The climate in truly equatorial towns like Singapore or Guayaquil may be languorous but is rarely stupefying. To experience heat that muddles the brain, the place to go is an inland locality at the fringe of

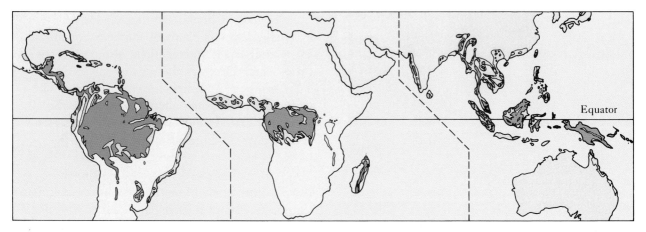

Present *(green)* and former *(yellow)* extent of humid tropical forest. Tropical dry forests once covered a similar area adjacent to the humid forest region, but these have been cleared to an even greater degree.

the tropics toward the end of the dry season. New Delhi, India or Asuncion, Paraguay will not be disappointing in this respect.

To find cooler temperatures, tropical residents go to higher elevations. As the atmospheric pressure decreases, the air expands and cools at the adiabatic lapse rate. In humid tropical mountains the lapse rate is roughly −0.55°C for every 100-meter increase in elevation. Accordingly, mean annual temperature at the equator drops below freezing at approximately 5000 meters, a level that coincides closely with the lower limit of permanent glaciers in the Andes.

Temperature per se limits forest vegetation within the tropical belt only at timberline, although it exerts many more subtle influences through its indirect effects on other components of climate. The intense insolation of tropical latitudes heats the lowland air, causing it to waft gently upslope into the mountains. As the air

rises, it cools, and eventually reaches the dew point. At this level a flat-bottomed layer of clouds hangs over the mountain slope from mid-morning to late afternoon. Since tropical weather systems are often driven by convection, rather than by large-scale wind systems such as the jet stream, such cloud banks tend to form at the same elevation with daily monotony.

These unobtrusive weather processes produce a sharp discontinuity in the vegetation of the mountainside. Below the cloud bank, the climate is essentially that of the lowlands, albeit somewhat cooler and wetter. In the clouds, the climate is one of almost perpetual gloom and dampness. As dense mists drift through the canopy, moisture condenses on every surface, producing a constant slow drip. The damp trunks and branches nurture mosses, ferns, and the nearly microscopic seeds of orchids, bromeliads, and other epiphytes. The profusion of plants in these cloud forests gives them a

Relics of the age of dinosaurs, tree ferns are a hallmark of cloud forest vegetation. Tree ferns reproduce as do all other ferns by microscopic spores that are nurtured by the perpetually moist environment.

Higher upslope, the vegetation on many mountains again changes character. The trees become stunted and gnarled, and on exposed ridges they may be no taller than a person. Their branches groan with multihued lichens, not the mosses of the cloud forest, and their leaves are minute, stiff, and tightly packed around upright twigs. This is the elfin forest, the final effort of woody plants to scale the heights. Perpetual chill and gloom, coupled with fully saturated air, stifle transpiration, the

uniquely exotic appearance, frequently enhanced by the presence of arching bamboos and umbrella-like tree ferns.

The drastic reduction in solar energy engendered by the daily cloud banks markedly reduces plant productivity. Cloud forests consequently grow slowly and fail to attain the stature of lowland forests. Although beautiful and fascinating to behold in the rare moments of sunshine, cloud forests tend to harbor few primates and other conspicuous mammals. However, they often do support an array of exotic birds that feed on fruit and nectar: cocks-of-the-rock, quetzals, and hummingbirds in the Andes; birds-of-paradise and honeyeaters in New Guinea.

Birds that feed on fruits, such as this Andean cock-of-the-rock, play increasingly important roles as seed dispersers as one progresses up the slopes of tropical mountains. Mammals, including primates, ungulates, and rodents, are the primary seed dispersers in lowland forests, although birds are still the essential dispersers of certain groups of plants.

evaporative loss of water from leaves. Reduced transpiration retards the upward flow of nutrient-containing sap, and growth slows almost to a standstill. Spindly twisted trees only a few meters tall and less than 10 centimeters in diameter can be hundreds of years old. In the Andes, this is home to spectacled bear and miniature pudus deer, and, in the volcanos of East Africa, to mountain gorilla.

Elfin forests have long been a subject of speculation among botanists, for they afford prime examples of the *Massenerhebung* effect. This term refers to the downslope displacement of vegetation zones on small, isolated mountains. On the more lofty tropical ranges, such as the Carstensz in Irian Jaya, Mount Kinabalu

The gnarled trees of this "elfin" forest grow at an imperceptible rate in the nearly sunless climate near timberline in the Peruvian Andes. Roiling mists condense on every available surface, nurturing carpets of mosses, lichens, and other epiphytes. In a world of eery silence, the stillness of the elfin forest is broken only by the faint lisp of a passing bird or the "peep" of a hidden frog.

in Borneo, the Ruwenzori in Africa, and the Andes in South America, elfin forests are typically met at elevations above 3000 meters. Botanists have thus been puzzled at finding physiognomically similar formations at only 1000 meters on isolated peaks, such as Pico del Este in Puerto Rico.

It is clearly not temperature that is critical to producing the *Massenerhebung* effect, for Pico del Este is no colder than other locations at the same elevation. The puzzle centers on whether the lowering of vegetation zones results from the suppression of photosynthetic activity under heavy cloud cover and high humidity, or from severe nutrient deprivation induced by the unrelenting precipitation and leaching (removal in solution) of mineral elements. Further study will be required to resolve the issue.

Timberline occurs at the point where net photosynthesis can no longer support the cost of maintaining a woody superstructure. In the tropics, this always lies well below the snow line, coming at around 3500 meters in New Guinea and the tropical Andes, although in both regions a few woody plants can be found in sheltered ravines even above 4000 meters. In a counterintuitive pattern, timberline actually increases in elevation away from the equator, reaching levels well above 4000 meters on some Mexican volcanos and in the Himalayas, before declining again farther northward. The higher timberline on these subtropical mountains is undoubtedly due to the longer, warmer days of the summer climate, which allows a more vigorous growing season. At 4000 meters on the equator, temperatures hover near or below freezing every night of the year, and plant growth is barely perceptible.

Vegetation and Rainfall

Rainfall, like temperature, tends to be high around the equator, because this is where the sun's evaporative power is at its highest. On average, precipitation decreases gradually away from the equator, tapering off rapidly near circum-global dry belts that exist at latitudes around 25 to 30 degrees in most parts of the globe. Rainfall is subject to much greater local variation than temperature, as a result of exposure to trade winds or proximity to oceanic currents or mountain ranges.

Just as mean temperature within the tropics is tied to elevation, so does annual rainfall

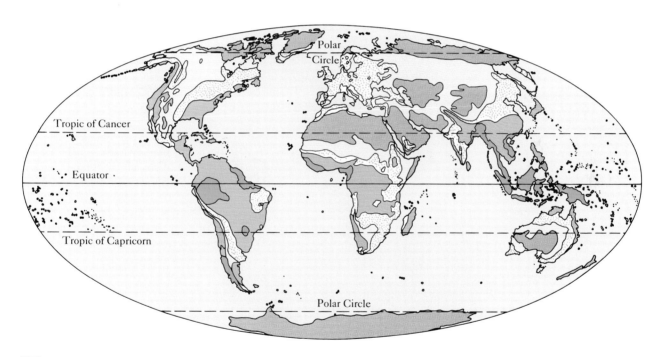

>2000 mm

1000–2000 mm

500–1000 mm

250–500 mm

<250 mm

Vegetation is closely tied to rainfall, especially in the lowland tropics where warm temperatures and strong sunlight encourage evaporation. Within the tropical belt, evergreen forests are the norm wherever the annual precipitation exceeds 2000 mm; savannahs and dry deciduous forests predominate toward the low end of the 1000 to 2000 mm range, while deciduous to partially deciduous forests prevail toward the upper end; between 500 and 1000 mm one can find low deciduous forests, thorn forests, dry savannahs, and semiarid grasslands; low, shrubby vegetation occupies most areas lying in the 250 to 500 mm range; vegetation is sparse or absent wherever the rainfall is less than 250 mm.

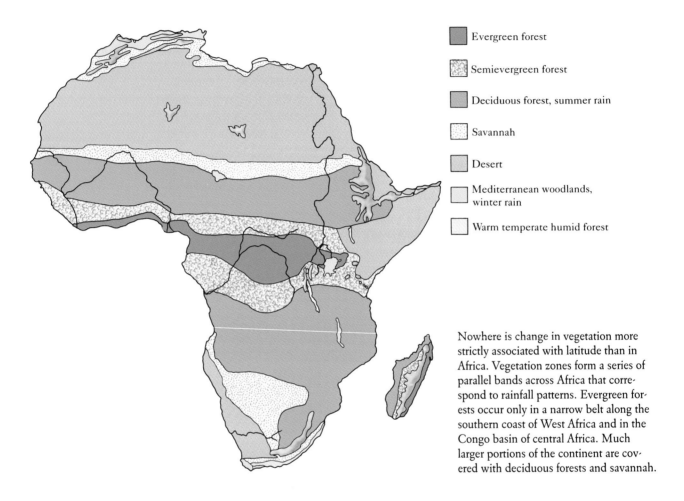

Evergreen forest

Semievergreen forest

Deciduous forest, summer rain

Savannah

Desert

Mediterranean woodlands, winter rain

Warm temperate humid forest

Nowhere is change in vegetation more strictly associated with latitude than in Africa. Vegetation zones form a series of parallel bands across Africa that correspond to rainfall patterns. Evergreen forests occur only in a narrow belt along the southern coast of West Africa and in the Congo basin of central Africa. Much larger portions of the continent are covered with deciduous forests and savannah.

depend on the length of the rainy season. In general, the farther away from the equator, the longer the dry season and the stronger the seasonal contrast. As rainfall decreases, the number of dry months in the annual cycle tends to increase. (For biological purposes, a dry month is defined as one in which precipitation fails to equal the evaporative potential of the climate.) The amount of rain falling in the wet months of a seasonally dry climate is often similar to

that recorded in an average month of a continuously wet climate. Regions within the tropics receiving low annual rainfall are thus better thought of as seasonal than as dry.

As the seasonal contrast intensifies away from the equator, forests show increasing degrees of leaf loss (deciduousness) in the dry season. Leaf drop in semideciduous tropical forests may be extensive only in years of unusually severe drought. In a normal year with 4 or

5 dry months, the crowns of many of the taller trees lose their leaves for varying periods, often only a few weeks, and leaf out again before the rains arrive. In such forests, the deciduous habit is mainly expressed in the canopy, as the stress of water shortage becomes accentuated at increasing heights. The understory tends to remain green throughout the year, although wilting may be apparent toward the end of the dry period.

Tropical deciduous forests are found in regions where the dry season persists for 5 to 7 months. In temperate forests, leaf drop coincides with the risk of subfreezing temperatures, but in tropical deciduous forests foliage often persists a month or two after the cessation of regular rains, drawing upon water stored in the soil. During the middle of the dry season, these forests may be as bare of leaves as the winter woods in Pennsylvania. Regreening often anticipates the return of heavy rains by one or two months, allowing trees to take advantage of the strong sunlight typical of the late dry season. A dry forest in full foliage can be as verdant and shady as a deciduous forest in temperate North America or Europe.

A lengthening dry season implies a reduced growing season, lowered productivity,

Tropical deciduous forests such as this one in Costa Rica can become as bare of leaves in the dry season as a temperate forest does in the winter. New foliage may sprout a month or more before the rains return in anticipation of a growing season that can be as brief as three months.

and an increased metabolic cost of maintaining living structures over the inactive period. These costs are especially severe at the elevated temperatures of the tropics, and may limit the height attained by trees. Tropical deciduous forests are accordingly low in stature and contain few trees more than 30 meters tall. At temperate latitudes, where dormancy is imposed by cold rather than drought, the physiological stress on trees is much less, and 30-meter trees are the norm in mature stands.

The Climatic Limits of Tropical Forest

In many parts of the world, tropical forests abruptly grade into savannah woodlands or grasslands. Such nonforest vegetation types are frequently maintained by natural wildfires, just as was the tall grass prairie of the midwestern United States before settlement. In more humid regions, secondary (human-induced) grasslands commonly develop after removal of the primary forest for shifting agriculture. These grasslands are perpetuated by the fires that local residents set to drive game or to stimulate new growth for livestock. Once large areas have been converted to grasslands in this way, it becomes extremely difficult to reestablish tree cover, even in climates that would otherwise support evergreen forest. The spread of secondary "alang-alang" (Imperata cylindrica) grasslands has been especially pronounced in parts of Southeast Asia, the Philippines, and New Guinea. This is a pernicious trend because alang-alang is a coarse, fibrous, and nutritionally barren grass, inedible even to goats.

Where human intervention is not a major factor, an annual rainfall of 1500 mm often coincides with the forest–savannah boundary. Forests may occur in still drier regions where fire is infrequent or where the soil remains moist throughout the dry period, such as in the galleries along stream courses. In certain areas, deep, fine-textured soils may support semideciduous forests on a rainfall barely over 1000 mm. Conversely, savannahs may appear in otherwise forested regions receiving more than 1500 mm precipitation on outcrops of coarse or shallow soils of low water holding capacity. Such climatically anomalous savannahs occur on the equator in central Gabon and in the heart of the Brazilian Amazon near Santarem.

Topography is often decisive in tipping the balance. It is commonplace to see fire-maintained grassland on the windward slope of a ridge and forest in the ravines and on the leeward slope. Where natural fire protection is afforded by topographically complex terrain, dry forests occasionally extend into climates producing even less than 1000 mm of rain, as in Piura on the northern fringes of the Peruvian coastal desert.

Tension between forest and nonforest vegetation can also be regulated by the action of large herbivores. A particularly notorious case is that of the Tsavo National Park of southeastern Kenya. According to legend, the region was a vast savannah in the mid-nineteenth century. The lure of ivory attracted professional hunters who decimated the elephant population. Within a few decades, an impenetrable thicket of thorny acacias had taken over the landscape.

The timberline on many tropical mountains is artificially lowered by annual burning of the alpine grassland. Here in the Peruvian Andes, retreat of the forest is evident in the brown foliage of trees killed by fire in the large burned patch occupying the center of the picture.

Elephants gradually recovered their numbers through the first half of the twentieth century, but not sufficiently to have a major impact on the vegetation. Then in the 1960s, large numbers of elephants migrated into the park, fleeing human encroachment in surrounding areas. An artificially swollen elephant population, in combination with a severe drought, led to an unsustainable onslaught on the vegetation. Hundreds of elephants starved, while those that survived reduced the acacia thickets and baobabs to shreds. Open grasslands once again appeared in the Tsavo region. Now, a century after the first ivory harvest, illegal hunting has again thrown elephant numbers into decline, and once more acacias have free reign to reclaim the land.

This anecdote illustrates the conditional nature of the climatic boundary between tree-dominated and grassland habitat. In general, a sparing but equitable distribution of moisture is more conducive to forest than is sharp seasonality; cloudiness can ameliorate the force of a dry season; soils, topography, and large herbivores can all play roles, as we have seen. The subject is complex, but important, as more and more of the land that once supported tropical forest is converted to nonforest vegetation.

The Latitudinal Limits of Tropical Forests

There are only a few places in the world where moist climates extend in a continuous belt from the tropics into the temperate zone. (Deserts interrupt the continuity of moist climates in most parts of the world.) The exceptional locations occur on the eastern margins of continents where trade winds blow across warm oceanic currents, picking up moisture that is wrung out over land. It is only in such spots that true tropical forests extend beyond the geographical limits of tropics: northeastern India at the foot of the Himalayas in Assam, southeastern China, northeastern Australia, southeastern Brazil in the state of São Paulo, and southeastern Madagascar. In each of these regions, narrow fingers of habitat remain warm enough and moist enough throughout the year to support tropical vegetation.

Unlike the situation in tropical mountains, where woody vegetation commonly extends upward into the frost zone, the latitudinal limits of tropical forest are closely associated with freezing temperatures. Evidence that frost limits the occurrence of tropical tree species was serendipitously derived from the great freeze that hit southern South America in July 1975. Housekeepers will recall it as the time when the price of coffee in the United States suddenly shot up over $7 a pound. The incident was a once-in-a-century event that did great damage, not only to Brazil's coffee crop but to natural vegetation in the state of São Paulo. Some alert botanists conducted a survey afterward and found that the greatest damage was suffered by plant species with tropical affinities, while many of those with south temperate affinities escaped unscathed. Rare events that are selective in their action thus may have decisive effects in regulating the composition and diversity of vegetation.

Because of the intervention of circumglobal dry belts, tropical and temperate forests come into contact only in the few restricted zones mentioned above. It is thus extremely unfortunate that several of these transition forests have already been obliterated by deforestation. The only one I have been privileged to see myself was on the slopes of the Sierra Madre del Sur in the Mexican state of Guerrero. One of the most beautiful and impressive forests I have seen anywhere, it was also botanically unique. The canopy was composed of towering trees—oaks, elms, walnuts, maples—belonging to familiar North American genera; the understory was formed almost entirely of tropical genera. My fascination for the vegetation was tempered by my frustration as every day I watched the logging trucks roll by.

The Classification of Tropical Forest Vegetation

As viewed from a low-flying aircraft, tropical forests reveal only hints of their infinite variety. Countless treetops merge into a roughly textured green carpet that extends to the horizon in all directions. The expansive scene is

An aerial view of an Amazonian whitewater river meandering through its broad floodplain. Erstwhile meanders, now sealed off by freshly deposited sediment, linger as oxbow lakes to the left and right of the present channel. Early successional vegetation advances toward the beach of the expanding meander loop in the center.

blocked into jigsaw puzzle sections by the stringy meanders of anonymous streams. Crowns of diverse heights, sizes, and colors contribute to the uneven texture, alluding to the great variety of trees that lie below. The ensemble is made up of every imaginable hue of green, best noticed upon close focusing, as when inspecting the detail of a pointillist canvas. One's eye is invariably drawn to the scattered singularities that punctuate an otherwise monotonous pattern: a leafless emergent crown, perhaps a feathery palm, and here and there an

arresting splash of color where a tree beckons its pollinators from afar.

The eye's tendency to perceive uniformity rather than detail in such a scene generates an illusion of sameness. This false perception paradoxically arises out of the very diversity it hides. Northern forests are often dominated by one or a few related species. Oaks, pines, and spruce are all distinct at a glance, and forests composed predominantly of one or the other are readily distinguished by superficial appearance. Our ability to discriminate on the basis of gestalt fails altogether when a forest is composed of several hundred trees, none of which is dominant. Another forest, perhaps only a few kilometers away, also containing several hundred species, will superficially look the same, although botanically it may be quite distinct. The very biological richness that makes these forests so intriguing scientifically has greatly impeded progress in classifying them.

In the north temperate region, forest types are distinguished on the basis of composition and named for the dominant species. In the eastern United States, for example, boy scouts and girl scouts learn to recognize oak-hickory, beech-maple, spruce-fir, and other familiar associations. European forests are classified to an even finer degree on more subtle compositional nuances. Over the reach of the 48 coterminous states, 135 natural plant formations have been characterized in a comprehensive system that is employed by conservation organizations to establish priorities for preserving North American biodiversity. It will be a long time before there is such a classification available for any comparably large area in the tropics.

Vegetation maps exist for many tropical countries, but they are based on much cruder divisions, generally ones that convey little or no botanical information. The starting point is often the basic distinction between forest and nonforest cover. In this first cut of the deck, "forest" refers to all vegetation supporting a continuous tree canopy; "nonforest" refers to everything else, including open woodlands, savannah, desert scrub, and alpine grassland. Most of the latter formations are associated with seasonal or arid lowland environments, or with cold high-elevation zones above timberline.

Under the heading of forest are many readily distinguishable types: most basically, primary and secondary forests. Primary forests are natural forests, sometimes referred to as "virgin" or old-growth forests. Although ancient human disturbance cannot be precluded, primary forests are characterized by an uneven age structure and the presence of at least some very old trees. In contrast, secondary forests spring up after some major disturbance (usually human-induced), and begin as even-aged stands. The species composition of secondary forests is often radically different from that of the primary forest that originally occupied the same land. Beyond the primary versus secondary dichotomy are other superficial distinctions, evergreen versus deciduous, for example, or flooded (occasionally or seasonally) versus upland (never flooded). These are still very coarse divisions, within which many variants can be discerned.

Nearly everywhere, finer divisions are recognized by local cultures that have evolved their own terminologies. Sometimes the indigenous classifications are detailed and perceptive, distinguishing subtle gradations in structure or species composition that reflect local variation in soils or disturbance history. The Kayapo Indians of Brazil, for example, have an elaborate vocabulary to describe stages in the cycle of regeneration of abandoned gardens as they return to forest.

Much of this subtlety vanishes in the largely one-way cultural interchange between indigenous and modern cultures. When settlers move into the forest, they often arrive with no foreknowledge of the environment and tend to scorn the indigenous people as savages from whom nothing of value can be learned. Thus, a refined classification system, such as that of the Kayapo, typically degenerates to a single term. In Peru, abandoned gardens revert to *purma;* in Brazil to *capoeira.* Knowledge acquired over millennia by the indigenous people has been irretrievably lost. Modern science has a long way to go in recreating this sophistication.

The primary impediments to progress in classifying tropical vegetation have been the low priority given by governments to the task, a scarcity of trained personnel, the difficulty of access to many regions, and the prodigious plant diversity of tropical forests, in which no species can be recognized as dominant. Few tropical countries can count even a handful of foresters capable of identifying trees in their primary forests. With so few resources to commit to the task, it seems likely that many tropical forests will vanish long before they have been classified and mapped.

Existing classificatory schemes, with few exceptions, are based on almost anything but tree species composition. Gross physiognomy, topographic criteria, phenological behavior, and climate have all been used. Each country or region has settled upon the system that is best suited to the prevalent conditions.

The vegetation map of Mexico, for example, employs a mixed system befitting the country's geographical location astride the transition from temperate to tropical. Plant formations at middle and high elevation that experience winter cold are dominated by such temperate genera as oaks, pines, and firs. Here the classification is of the temperate type, based on dominant species or genera.

The tropical lowlands are accorded an entirely different treatment. In the absence of for-

mations dominated by single taxa, the classification is based on the degree to which the tree canopy loses its foliage in the dry season. The leafing behavior of the canopy, in turn, reflects the seasonality of rainfall. The wettest areas support tropical evergreen forest; areas experiencing mild seasonal drought are semideciduous; markedly seasonal zones, especially along the west coast, are designated as deciduous. The divisions correspond only roughly to compositional units, as the diagnostic criteria are physiological rather than taxonomic.

Costa Rica, Panama, and several other countries in Central America and northern South America have adopted the "Holdridge system," named in honor of its originator, a U.S. expatriate and pioneer tropical forester who has made his home in Costa Rica. Holdridge found that the occurrence of plant formations in Costa Rica is strongly controlled by climate, with both moisture and temperature (elevation) exerting major influences. Accordingly, he devised an elaborate scheme for classifying climates based on mean annual temperature and rainfall. The arbitrary divisions were then assigned names designating a particular type of vegetation, for example, tropical dry forest or premontane wet forest. As a heuristic device, the Holdridge scheme has had considerable practical value in Central America, where, by and large, plant composition and climate show strong and consistent associations. It has not been so effective elsewhere, as in Amazonia, where climate is uniform over vast areas, and vegetation is affected by unrelated environmental factors.

In Peru, for example, foresters are confronted with a topographically complex land-

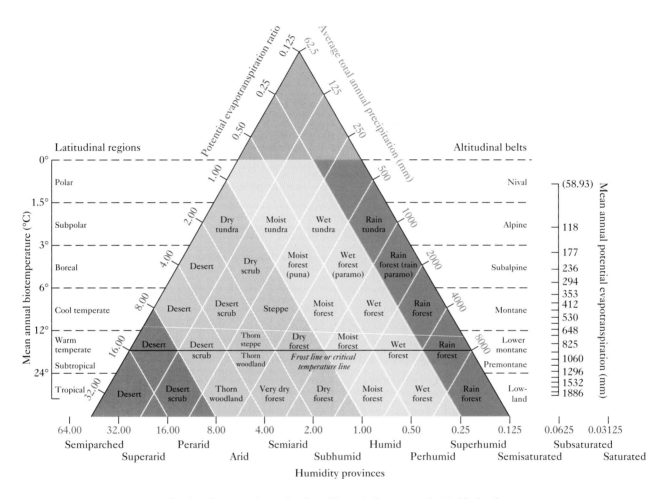

The classification scheme developed by U.S. forester Leslie Holdridge for vegetation in Central America. Rainfall increases from left to right on a declining axis; temperature decreases from bottom to top. The maximum evaporation plus the transpiration that could occur in a given environment is termed the potential evapotranspiration. Where rainfall exceeds the potential evapotranspiration, the ratio of the two quantities is less than 1.0, and evergreen vegetation is predominant. Drier climates are represented by ratios of greater than 1.0. Within Central America the cells in the diagram correspond reasonably well to recognized vegetation formations.

scape and an overwhelming diversity of tree species, many of which do not yet have scientific names. In the Amazonian lowlands, large areas are subject to inundation, and these areas support vegetation formations distinct from those found in the uplands *(tierra firme)*. In the headwaters region along the eastern base of the Andes, broken terrain and varied geology create an even more complex situation. That much of the region is roadless and scientifically unexplored only adds to the difficulty of mapping vegetation.

Given these handicaps, Peruvian foresters have classified vegetation on the basis of topography, distinguishing such categories as permanently inundated forest, seasonally inundated forest, upland forest, and hill forest. Within each of these designations, there might be a score of compositionally distinct plant communities. Nevertheless, the system has practical value in planning the location of roads and in estimating the costs of timber extraction.

The lower Amazonian region in Brazil presents yet another set of conditions. Relief is relatively insignificant, as mountains are all but lacking. Instead, rivers dominate. Streams arising in the Andes to the west carry huge loads of silt. Having a characteristic light brown color, they are termed "whitewater" rivers, perhaps because they give rise to blinding reflections. Whitewater rivers deposit their loads of silt at the foot of the mountains, building up broad floodplains. The friable, fine-grained sediment of these floodplains is both highly fertile and easily eroded. Channels constantly shift, giving rise to sinuous meanders and oxbow lakes. These floodplains, some of them more

than 10 kilometers across, would have been converted to agriculture centuries ago were it not for seasonal inundations that may persist for half the year. During flood season, the water may rise 10 meters or more, confining local populations to their floating or stilt-legged houses. The challenge imposed by these inundations to the establishment and growth of trees has led to the evolution of a distinct "varzea" forest, composed largely of characteristic species.

Another major Amazonian environment is created by rivers draining the Precambrian sandstones of the Guiana Shield. This is one of the oldest geological formations on earth, having emerged from the sea long before the origin of multicellular life. Occupying a vast region in the northeastern quadrant of South America, the Guiana Shield is characterized by leached and weathered soils that are among the poorest on earth. The water that drains out of them is almost devoid of mineral content. To conserve scarce nutrients, many plants of this region protect their foliage with high concentrations of tannins, acidic compounds that are difficult for herbivores to digest. The tannins leech into the groundwater after the leaves have fallen. The outflow of the Rio Negro and other major affluents of this region, although clear, is deeply stained by tannins, and hence referred to as "blackwater."

The Rio Negro and other blackwater rivers are flanked by inundation zones, but the lack of sediment and the acidic sterility of the water pose conditions quite distinct from those in the whitewater varzea formation. The analogous "igapo" formation of blackwater rivers

Sunlight reflected from a sandbar shimmers through the tea-colored waters of the Rio Negro in Brazil. Water flowing over the bar generates wavelike undulations in the white sand. The dark stain is produced by tannins that leach out of decomposing leaves in the swampy, infertile drainage basin. Blackwater rivers carry little sediment and meander less than whitewater rivers.

constitutes a second distinct assemblage of inundation-tolerant trees.

Peru and Brazil, although sharing a 1500-kilometer border, have evolved entirely different systems for classifying their forests. In Peru, topography provides an appropriate criterion in the hilly country at the base of the Andes, while in Brazil the great river plains and the geology of their basins have been seen as more relevant. The common denominator that unites the two schemes, as well as the others described above, is that they are based on practical expediency, guided by the intuition of experienced foresters.

Such preliminary efforts to classify tropical vegetation are a necessary and valuable first step, but they fall far short of standards we take for granted in the temperate world. Unfortunately, the goal of preserving tropical diversity requires detailed information on the composition of biological communities, a requirement that cannot be satisfied by the current state of knowledge.

Although the examples above are typical, strides toward the goal of composition-based vegetation mapping have been made in a few countries, among them, Costa Rica, Venezuela, Malaysia, and Indonesia. British and Dutch foresters expended major efforts in the latter two countries, both before and after World War II. The interest of the colonial powers was inspired by the fact that the primary lowland forests of Southeast Asia are the best stocked and most valuable of any in the tropics. The exceptional commercial value of these forests is attributable to a prevalence of trees in the plant family Dipterocarpaceae. Dipterocarps tend to be ramrod straight, up to 60 meters tall, and to possess excellent working properties. Many species are collectively sold under the name of "Philippine mahogany," the term being a deliberate misnomer designed to promote market value.

Dipterocarps in Southeast Asia play a role not unlike that of oaks in the forests of the southeastern United States. There are some 30 species of oaks in this section of the country, many of which are closely associated with particular soil conditions. One finds white and black oaks on well-drained upland clay soils, shingle oaks on calcareous soils, water and willow oaks in moist riverbottoms, turkey and blackjack oaks on dry sandhills, and live and laurel oaks in maritime forests. Similarly, in Southeast Asia the two or three most abundant species of dipterocarps serve to characterize the floristic composition of many stands and to convey information about the underlying substrate. There are hundreds of dipterocarp species in Southeast Asia, however, so that the number of recognizable associations greatly exceeds that of oak forests in the southeastern United States.

A Role for Remote Sensing

In the future it is inevitable that most large-scale mapping of tropical forests will be conducted by means of remote sensing, either from satellites or aircraft. At this writing, efforts to classify natural vegetation formations from remote sensing data are at a preliminary stage. Yet, both the resolution of images and the computerized processing of data are improving rapidly, suggesting that powerful new methods will become available within the decade.

Two technological approaches seem especially promising. One is the analysis of visible

Two dipterocarps tower over lesser trees in a Malaysian forest. Giants such as these are rapidly becoming a thing of the past, as logging and deforestation sweep across the land.

and infrared light reflected from vegetation. The receptors aboard satellites are rather analogous to the rods in the human eye, in that they are sensitive to different colors of light. Interpretation of the information received by such sensors may be simple in the case of water (blue) or ice (white), but is far more difficult when subtly varying shades of green are involved. With current technology it is relatively easy to distinguish the dark green of a conifer forest from the light green of a deciduous forest, a distinction easily made with the naked eye from a high flying airplane. Discriminating different types of dipterocarp forest, however, is a far more challenging proposition, and a task lying well beyond present capabilities. It is not yet certain whether the reflectance approach is intrinsically capable of meeting the challenge.

If not, there is another technology on the horizon, one that is just now in the preliminary stages of development at the Jet Propulsion Laboratory in Pasadena, California. It is a form of radar that beams angled, "side-scanning" pulses in the centimeter band from an aircraft flying at 10,000 meters, a normal height for long distance jet travel. Maximum reflections are derived from objects whose dimensions approximate the wavelength of the beam. Three wavelengths are employed simultaneously to obtain information about the presence of objects in the size ranges of leaves, branches, and trunks. The multiband reflection thus carries information, not only about the canopy but about structures below the canopy as well. The capacity of radar to "see" through the superficial foliage gives it a resolving power potentially far superior to that of systems based on visible light. Preliminary evidence from trials conducted in Belize indicates that this system is capable of discriminating all the vegetation formations recognized by a botanical expert. This appears to represent a major breakthrough, because no prior technology has been able to achieve such fine levels of discrimination.

The ability of remote sensing to cover large areas quickly will inevitably make it the method of choice for vegetation mapping in the future. Even if the vegetation units discriminated by such indirect methods do not correspond precisely to the ones that would be recognized by botanists, the prospect is for maps that are far better than any that exist at present. A biologist can't fail to see irony in the United States' having spent billions developing the technology to read license plates from space, while the state of vegetation mapping in Amazonia and elsewhere in the tropics remains in the scientific stone age.

The special scientific value of tropical forests is that they offer our last chance to study nature in its prehistoric condition, nature as it evolved over eons past. The bustling, industrialized temperate regions, with their superhighway grids, acidic atmosphere, and flood-controlled waterways are a far cry from nature in its pristine form. Over much of North America, even within some of the most acclaimed national parks, ecosystems have become simplified and unbalanced by a lack of large predators and other key species that were present when the Pilgrims landed. Later we shall see that such creatures play critical roles in maintaining the diversity and integrity of ecosystems.

The few remaining wilderness regions of the tropics offer science a last chance to understand diversity, why there are so many species, how they form, and how they coexist in stable communities. Pristine ecosystems still exist in parts of South America, central Africa, Indonesia, New Guinea, and some other Pacific islands. These ecosystems are priceless and irreplaceable assets, for they constitute some of the few remaining controls for biological science, the yardsticks against which to gauge humanity's impact on the environment.

If, as seems likely, we lose the tropical forest in another 30 years or so, then science will be irrevocably crippled. Trying to understand the working of evolution then would be like trying to deduce the functioning of a stereo if one had only the component parts and had lost the assembly manual. It might be possible to reverse engineer each component separately and deduce its operation, but the emergent properties of the whole would remain an unresolvable mystery.

2

The Paradox of Tropical Luxuriance

"The primeval forests of the equatorial zone are grand and overwhelming by their vast-ness and by the display of a force of development and vigour of growth rarely or never witnessed in temperate climates." Thus the renowned nine-teenth-century naturalist Alfred Russel Wallace described the profusion of foliage and the im-mense trunks of a tropical forest.

Impressed by the luxuriance of the rain for-est, many visitors from the temperate zone have concluded that the soil beneath the towering trees must be richly endowed with the nutrients plants need to grow. But recent research has shown that this seemingly straightforward inference is based

The recycling of nutrients essential for plant growth is assisted by these *Hygrocybe* fungi growing on the floor of a Malaysian forest.

Net Productivity of Vegetation in Ecosystems of the World

Type of vegetation	Amount of dry matter*	
	Normal range	Mean
Tropical rain forest	1000–3500	2200
Tropical seasonal forest	1000–2500	1600
Temperate forest		
Evergreen	600–2500	1300
Deciduous	600–2500	1200
Boreal forest	400–2000	800
Woodland and shrubland	250–1200	700
Savannah	200–2000	900
Temperate grassland	200–1500	600
Tundra and alpine	10–400	140
Desert and semidesert scrub	10–250	90
Extreme desert	0–10	3

*Measured in number of grams per square meter per year.

on an improper analogy with temperate forests. Farmers familiar with the middle latitudes know from pragmatic experience that lofty trees are good indicators of rich and fertile terrain. Yet, paradoxically, this bit of conventional wisdom does not transfer to the tropics, where the soils are often barren of important nutrients.

In one sense, however, the perception is correct. Tropical forests growing on soils of average quality are highly productive, exceeding other types of terrestrial vegetation in their scale of photosynthetic activity. Productivity can be estimated by collecting, drying, and weighing the mass of leaves, fruits, and branches that fall to the forest floor over a yearly cycle. This "litter" represents about 40 percent of the aboveground productivity of temperate forests, and a somewhat higher fraction of that in tropical forests. Comparing the values obtained for a wide range of ecosystems, the mean for tropical evergreen forests is found to be almost twice that of temperate forests. Why then is it necessary to withhold one's judgment about the soils tropical forests grow on?

Nutrients and Soil

The ingredients of plant growth are simple and nearly ubiquitous: carbon dioxide and oxygen from the air, water from rainfall and the ground, and some 13 essential mineral elements from the soil. Two major groups of essential elements are commonly recognized: macronutrients, such as nitrogen, phosphorus, potassium, calcium, and magnesium, which are required in relatively large amounts, and micronutrients, such as boron, cobalt, molybdenum, and others, which are required in trace amounts. In most tropical soils, one or more macronutrients (most often phosphorus) are present in amounts limited enough to restrain growth.

Ironically, the so-called mineral soil that underlies many tropical forests is nearly devoid of soluble minerals that can be absorbed by roots. The sterility of these soils results from exposure to torrential rains over millennia. As mildly acidic rainwater percolates downward through the soil, all soluble mineral elements are washed away. This leaching process leaves behind a barren substrate composed of the most insoluble materials contained in the parent rock. This residue is what we call soil.

Soil is formed in a complex process that involves both chemical and biological activities. Weathering of the bedrock is the first step. In this process, chemical reactions promoted by heat and moisture gradually disassociate the parent rock, generating fine particles whose high surface-to-volume ratio favors release of mineral elements into solution. At the same time, decomposing plant and animal remains contribute to the accumulation of organic matter at the surface. The typically stratified soil profile that results has a highly organic layer on top (topsoil), underlain by "mineral" soil (subsoil).

In most environments the weathering of bedrock is measured in millimeters per thousand years. Weathering is particularly rapid in the humid tropics because the accompanying chemical transformations are accelerated by high temperatures and copious percolation of rainwater. Rapid weathering does not mean that the released minerals are necessarily available to plants, however. Many meters of soil may accumulate above the bedrock, and roots will no longer be able to reach the minerals brought into solution by weathering deep in the soil. These minerals move out of the deep soil layers in the flow of groundwater, eventually seeping into streams that will carry the minerals away. Scientists are therefore able to judge the rate of weathering in the upstream watershed by measuring the mineral content of streams. As a consequence of rapid weathering, the humid tropical zone (broadly defined to include about 25 percent of the earth's surface) supplies about 65 percent of the dissolved silica (the most abundant constituent of rock) and 38 percent of the nonsilica minerals delivered by rivers to

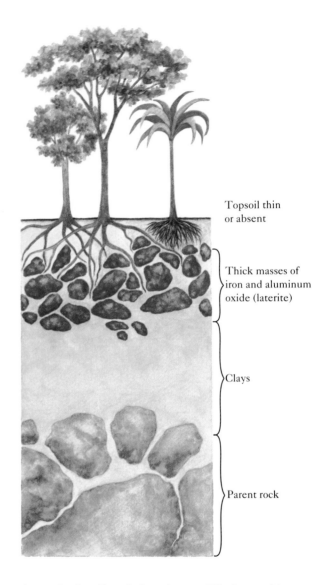

Topsoil thin or absent

Thick masses of iron and aluminum oxide (laterite)

Clays

Parent rock

A typical soil profile in the humid tropics. Weathering of the parent rock over millions of years has produced a deep bed of soil. In contrast, the layer of topsoil is thin because in the warm, moist climate organic matter decomposes too rapidly to accumulate. Hard concretions often form below the surface. Tree roots are confined to the superficial layers from which nutrients are captured as they percolate downward.

The Major Soils of the Tropics

Soil	Area (millions of hectares)	Percentage of area
Very low fertility		
Spodosols	19	1.3
Psamments	90	6.0
Low fertility		
Lithic (shallow)	72	4.8
Histosols (organic)	27	1.8
Moderate to very low fertility		
Oxisols	525	35.3
Ultisols	413	27.7
Variable fertility		
Aquepts	120	8.1
Moderate fertility		
Alfisols	53	3.6
Tropepts	94	6.3
Andepts	12	0.8
Mollisols	7	0.5
Fluvents	50	3.4
Vertisols	5	0.3
Other	2	0.1
Total	1489	

Included are sites with a mean annual temperature higher than 22°C, annual precipitation higher than 1500 mm, and a dry season of less than 4 months per year.

the ocean. The same region contributes about 50 percent of the world sediment load. These figures reveal the vigor of tropical weathering.

Where average temperatures are lower, as at high latitudes, or where rainfall is scant, as in the seasonally dry zones that fringe the humid tropical belt, weathering is less rapid, but for different reasons. Even though at temperate latitudes rock may be fractured by the action of frost and the breakup of the parent material accelerated, the rate of leaching is slower and the mineral elements are less soluble in the low temperatures.

In the seasonal tropics and subtropics, it is not low temperature but the alternation of wet and dry seasons that limits weathering. During the rainy season, the soil can become saturated and water percolates downward. The result is severe leaching of the kind that continues year round in the everwet portions of the equatorial zone. But when the rains cease, the soil begins to dry out, massively aided by the transpiration of plants. As roots withdraw more and more of the water stored in the soil column, there is a partially compensatory upward flow from deeper levels as a result of capillary action. The reverse flow carries with it freshly dissolved minerals from the weathering bedrock below. Thus, soils of the seasonally dry tropics are often more fertile than those of wetter regions.

The great diversity of geology and climate over the vast expanse of the tropics leads to a corresponding diversity of soils. If we look at this diversity on a global scale, we find that nearly two-thirds of all tropical soils are classified as oxisols and ultisols, two types of soils containing clays with scant content of soluble minerals. The moderate to strong acidity of these abundant soil types interferes with the ability of roots to take up nutrients. They range from mildly to severely infertile. About 7 percent of tropical soils are spodosols and psamments, soils of the lowest nutrient status, derived, respectively, from sandy alluvial terraces and highly weathered uplands. Their ability to support agriculture is virtually nil. The same can be said of lithic soils and histosols, the

Grasslike pandans appear to grow downward as well as upward in the reflection from this blackwater stream in Indonesian Borneo. The dark, acidic water is nearly devoid of nutrients, indicating that weathering in the upstream watershed is extremely slow. Attempts to promote agricultural development in such regions have been uniformly unsuccessful, and the resulting injuries to the environment may require centuries to heal. Blackwater basins are best managed by the devices of nature.

former consisting of outcroppings of bedrock and the latter of peat swamps. Altogether, these six types make up over three-quarters of the soils of the humid tropics.

Only about 20 percent of the soils of the humid tropical region are capable of sustaining agriculture with current technology, and most of this terrain is already intensively developed. Examples are the fertile volcanic highlands of Central America, the Philippines, Indonesia,

and Cameroon (alfisols), and the rich alluvial plains of the Ganges and Mekong in Asia (fluvents and aquepts). The practice of establishing agricultural experiment stations on soils such as these has created a misleading impression of the agricultural potential of the tropics generally.

Where extensive forests still persist, as in Amazonia, the Guyanas, Central Africa, Borneo, and New Guinea, their very existence suggests the land is poorly suited for farming. Peasant farmers and indigenous agriculturists are expert at distinguishing good soil from bad. Everywhere, the worst is left until last. Had the exuberant vegetation of the humid tropics truly been a mark of fertile lands, as many casual travelers from the temperate world had imagined, it is nearly inconceivable that there would be any significant amount of rain forest left in today's overcrowded world.

Nutrient Cycling in Tropical Forests

How can the high measured productivity of tropical forests be supported by soils of low fertility? The answer to this question was not firmly established until the 1960s and 1970s, when scientists began to measure the chemical content of tropical vegetation and compare it to that of the underlying soils. It was then discovered that the nutrients of tropical ecosystems are mainly to be found in living and recently dead organic matter—in the plants themselves and in the litter of decomposing plant parts that carpets the soil.

That more nutrients are often contained in plant matter than in the soil suggests that plants are recapturing dissolved minerals as they are released during decomposition. In many tropical forests, the mineral soil serves mainly to anchor trees and supply water to their roots. Nutrients are concentrated principally in the thin cap of topsoil, where the final stages in the decomposition of plant and animal matter take place. Termites play a particularly important role in digesting the several tons of wood that fall to the ground each year in a typical hectare of primary forest. Bacteria, fungi, and myriad invertebrates participate in the release of nutrients in the soluble forms required by roots.

Even more vital to plant life are mycorrhizal fungi. These ubiquitous soil organisms provide the primary mechanism for the capture of nutrients used by tropical forest trees. The fungus invades tree roots and obtains nourishment by tapping into the host's vascular system. While the tree is obliged to feed the fungus, it benefits from the enormous nutrient-gathering ability of the mycorrhizal mycelium, a mat of microscopic fungal strands that effectively multiplies the surface area over which the tree is in contact with the soil. Many trees are so dependent on their mycorrhizal associates that they languish or die without them.

The efficiency of nutrient recovery has been confirmed by following the fate of isotopes introduced at the soil surface. In an often-cited experiment, more than 99 percent of the radioactivity applied to one Amazonian forest as calcium-45 and phosphorus-35 was retained in the root mat. More typically, recycling is able to recover 60 to 80 percent of the stocks of most nutrients in tropical forests. The experiment demonstrates that much of the nutrient stock of the tropics is caught in a cycle that takes nutrients from living plant to organic matter back to living plant. Because nutrient cycles are not 100 percent efficient, however, the remaining 20 to 40 percent of the stock must come from rainfall and the soil.

Meager though the nutrient stocks of the soil may be, they are a reservoir that the forest must draw on to supplement the nutrients

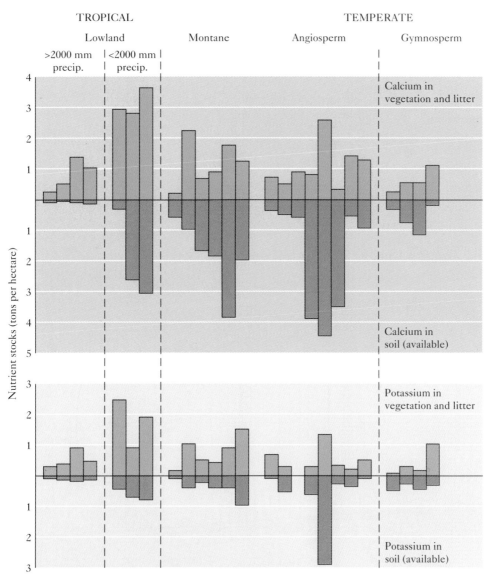

Distribution of calcium and potassium above *(green bars)* and below the ground *(brown bars)* in a number of forest ecosystems. The aboveground fraction represents minerals contained in living and decomposing plant tissues; the belowground fraction represents minerals contained in the soil. Note that these two minerals are relatively scarce in lowland tropical forests and that the stocks are largely contained in the aboveground fraction. Tropical dry and montane forests tend to be better endowed with calcium and potassium, both above and below ground, as are deciduous and coniferous temperate forests.

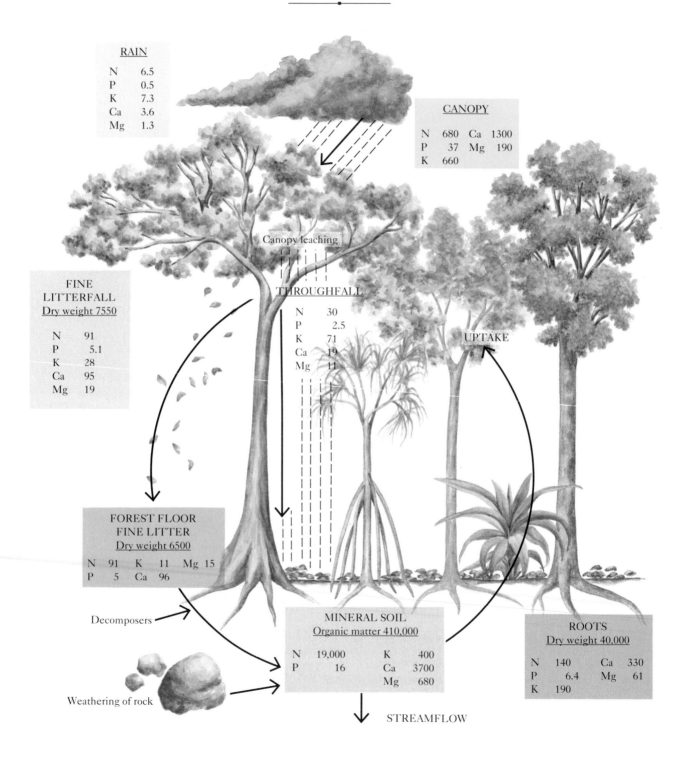

RAIN

N	6.5
P	0.5
K	7.3
Ca	3.6
Mg	1.3

CANOPY

N	680	Ca	1300
P	37	Mg	190
K	660		

Canopy leaching

THROUGHFALL

N	30
P	2.5
K	71
Ca	19
Mg	11

UPTAKE

FINE LITTERFALL
Dry weight 7550

N	91
P	5.1
K	28
Ca	95
Mg	19

FOREST FLOOR
FINE LITTER
Dry weight 6500

N	91	K	11	Mg	15
P	5	Ca	96		

Decomposers

Weathering of rock

MINERAL SOIL
Organic matter 410,000

N	19,000	K	400
P	16	Ca	3700
		Mg	680

ROOTS
Dry weight 40,000

N	140	Ca	330
P	6.4	Mg	61
K	190		

STREAMFLOW

bound up in living matter. These stocks would soon be depleted if nutrients lost to leaching were not replenished from additional sources. The most important of these are weathering and the atmosphere, which contributes nutrients as dust particles, pollen, and minerals dissolved in rainwater. Because few nutrients are supplied directly through weathering, the nutrient cycles of many tropical forests are extremely fragile. The survival of these forests depends on highly efficient mechanisms for recapturing nutrients from the stocks sequestered at the surface in living and decomposing organic matter.

Any large-scale disturbance, such as clearing or burning, that results in the breaking of recycling mechanisms will lead to a rapid loss of the accumulated nutrient capital. Once the nutrients are lost, the vegetation may not recover its former diversity or stature. New stocks can be accumulated only by slow release from weathering and by the import of dust and dissolved minerals in rainwater. In inherently poor soils, this process may take decades, even centuries. The regrowth that appears after disturbance will generally not contain the same species

Possessed of a pungent odor, the durian is at once Asia's most beloved and most despised wild fruit. The experience of savoring a ripe durian has been likened to that of eating Limburger cheese in an outhouse. Nevertheless, many city residents of the region have acquired a taste for the fruit and are willing to pay a handsome price for it.

(Opposite page) Nutrient cycling in a New Guinea montane rain forest. Nutrients enter the cycle through weathering of soil minerals and through rainfall and they leave by leaching into the underground water table. Nutrients contained in living plant tissues are returned to the soil from rain that filters through the canopy (throughfall) and from decomposing leaf litter and roots. Live roots then recapture nutrients from the soil and return them to the growing portions of plants. Note that the amounts of nutrients entering the cycle in rain are tiny compared to the amounts contained in the trunks, crowns, and roots of plants.

as the primary forest. Local extinction of many species is thus the inevitable consequence of using the land for short-term gain.

When human beings gather products from the tropical forest—whether timber, fruits, nuts, or game—we must stay within the limits imposed by the natural processes of weathering and nutrient capture. Forest products cannot be harvested indefinitely except at levels that do not exceed the natural rate of nutrient replenishment. Outputs and inputs must balance; this is the essence of sustainability. Here we confront a law of nature that we are reluctant to admit even in the industrialized countries, where farming practices routinely create rates of soil erosion that greatly exceed the rates of formation.

Ecosystem Characteristics in Tropical Moist Forests and Rain Forests

Parameter	Amazon caatinga San Carlos, Venezuela	Oxisol forest San Carlos, Venezuela	Lower montane rain forest El Verde, Puerto Rico	Evergreen forest Banco, Ivory Coast	Dipterocarp forest Pasoh, Malaysia	Lowland rain forest La Selva, Costa Rica	Moist forest, Panama
1. Total calcium in soil (kilograms per hectare)	195	7	176	—	115	6530	22,166
2. Total nitrogen in soil (kilograms per hectare)	785	1697	—	6500	6752	20,000	—
3. Total phosphorus in soil (kilograms in soil)	36	243	—	600	44	7000	23
4. Root biomass (tons per hectare)	132	56	72.3	49	20.5	14.4	11.2
5. Aboveground biomass (tons per hectare)	268	264	228	513	475	382	326
6. Root-to-shoot ratio	0.49	0.21	0.32	0.10	0.04	0.04	0.03
7. Percentage of roots in superficial root mat	26	20	~0	~0	~0	~0	—
8. Specific leaf area (square centimeters per gram)	47	65	61	—	88	139	131–187
9. Leaf area index	5.1	6.4	6.6	—	7.3	—	10.6–22.4
10. Predicted turnover time of leaves (years)	2.2	1.7	2.0	—	1.3	—	0.9
11. Leaf decomposition, k	0.76	0.52	2.74	3.3	3.3	3.47	3.2

Adaptations for Conserving Nutrients

Tropical forests are able to achieve impressive stature by efficient uptake and tenacious retention of scarce mineral elements. In some environments nutrients are at such a premium that plants have evolved a whole series of adaptations to concentrate and conserve nutrients. Nutrient cycling is not equally efficient in all tropical forests, however, because some soils are inherently more fertile than others. It does not pay, in an evolutionary sense, to invest heavily in adaptations designed to capture and conserve nutrients unless the investment is justified by superior growth performance. By this logic, the poorest sites should be expected to show the most conspicuous adaptations. This is clearly seen in a comparison of plant characteristics at a series of seven sites listed in the table, ordered from the least fertile on the left to the most fertile on the right.

Although the sites represent a range of several orders of magnitude in the availability of nutrients in the soil, there is surprisingly little variation in the aboveground biomass of the forests (the total mass of living tissue per unit area). The least fertile site, the San Carlos caatinga, supports a biomass only 20 percent less than that of the richest (Panama). The forest of highest biomass (Ivory Coast) occurs on a soil of intermediate fertility. Forests show obvious stunting only under the most extreme conditions of nutrient scarcity. Over the seven sites of this comparison, the lack of any clear relation between forest biomass and soil fertility provides a strong affirmation of the prevalence of efficient nutrient recycling. The greater the scarcity of nutrients, the more trees invest in nutrient capture and retention, so that over time each forest accumulates adequate stocks to support a forest of normal stature. Here then is why the appearance of tropical forests was so deceptive to the early European explorers, who extolled their luxuriance to eager listeners back home. Unless one has a keen and discriminating eye, it is not possible to guess the quality of the soil merely by looking at the trees.

A far better indicator of soil quality lies hidden beneath the surface. While the aboveground biomass of these forests varies within a factor of 2.3, the root biomass varies by a factor of nearly 12. Now there is a correspondence between soil quality and the measured parameter, because the poorest site has the highest root biomass and the richest site the lowest. Here is a striking illustration of the adaptive allocation of resources. Where the soil is fertile, plants compete more severely for light and invest heavily in their trunks, branches, and foliage; where the soil is poor and growth is limited by the capture of nutrients, the investment in roots must be relatively greater. In fact, the ratio of root biomass to shoot biomass is 16 times greater in the San Carlos caatinga than at the Panama site.

Moreover, at both San Carlos sites there is a conspicuous superficial root mat that is lacking at all the other localities. The competition for a meager supply of nutrients is so intense in the nearly sterile soil that the plants produce roots on the surface. These infiltrate the leaf litter, capturing nutrients as they are released during decomposition and as they enter through precipitation and throughfall. (The latter term refers to rainwater that has trickled through the canopy before reaching the ground.) Superficial root mats are a feature of the poorest sites, offering a better indication of soil quality than the size of the trees.

A root mat stands out in bas-relief atop the barren, sandy spodosol of this keranga forest in Sarawak, Borneo. Nutrients are so scarce in this environment that plants invest heavily in roots, extending them over the surface to capture mineral elements at the moment of release from decomposing litter.

Additional adaptations are found in the characteristics of foliage. Where soil fertility limits growth, trees must allocate more resources to constructing roots, as we have already noted. Shoot growth is consequently slower, and individual leaves can be retained longer before new foliage puts them in the shade. But leaves that are held longer are exposed to greater risk of being eaten by folivorous (leaf-eating) insects. Since the carbon compounds that are the immediate products of photosynthesis are more readily available than the nitrogen and phosphorus needed to build new cells, it costs the plant relatively little to reinforce its leaves with woody fibers, and to protect them against insects with tannins. The tough, thick, toxic leaves then resist decay once they have fallen to the ground.

The combination of long leaf-retention times and slow decay results in a greatly protracted recycling time. The whole system, as it were, runs more slowly as the rate of cycling is held back by a negative feedback loop. But how is a rice farmer from Java supposed to know this when he is resettled in Borneo, or an Andean peasant when he joins a program to colonize the Amazon?

Interactions of Soil and Moisture: The Vegetation Mosaic

In the distinct characteristics of forests at sites differing in soil quality, we can see a major evolutionary rationale for plant diversity. A tree that is genetically programmed to allocate resources to roots and shoots in certain propor-

tions, and to produce new leaves on a certain timetable, will perform best on soils offering a particular range of nutrient levels. Competition will ensure that other species will be more successful on richer or poorer soils. Accordingly, vegetation tends to show marked responses to geology, whether in temperate or tropical regions.

It is not merely geology that patterns a landscape, but topography as well. Small-scale topography, such as the gentle ups and downs in a rolling plain, influences vegetation primarily by affecting the availability of water. Where the water table lies close to the surface, the soil may become waterlogged. Oxygen diffuses very slowly through waterlogged soil, and roots become stunted or die from anoxia. A sudden rise in the water level, such as behind a beaver dam, will often kill trees for this reason. Where soils are porous and the water table lies deep, severe moisture stress can develop during dry periods. In extreme situations, lack of moisture can produce stunting reminiscent of that seen in desert shrubs.

Geology and topography interact to produce what can be called the vegetation mosaic, a patchwork across the landscape of distinct types of vegetation. An excellent example of a tropical vegetation mosaic is provided by the work of Carl Jordan and his associates at the University of Georgia. For many years they studied the region around San Carlos del Rio Negro in southern Amazonas Territory, Venezuela. Located on the Guiana Shield, this site is at the low end of the soil fertility spectrum. The plants at San Carlos accordingly possess adaptations that allow them to grow and compete in an environment in which mineral nutrients are at a premium.

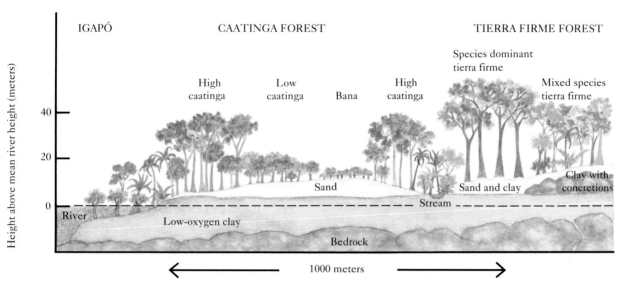

 labels:

IGAPÓ   CAATINGA FOREST   TIERRA FIRME FOREST

Species dominant
tierra firme

High
caatinga

Low
caatinga

Bana

High
caatinga

Mixed species
tierra firme

Height above mean river height (meters)

40

20

0

Clay with
concretions

Sand

Stream

Sand and clay

River

Low-oxygen clay

Bedrock

1000 meters

A schematic cross section of the varied terrain near San Carlos del Rio Negro, Venezuela, showing how dramatically the vegetation responds to small differences in soil type and elevation above the water table.

Viewed from the air, the terrain seems devoid of relief, but an observer on the ground notices slight undulations that differentiate well-drained uplands from stream courses and fringing swamps. The total relief, about 20 vertical meters, is only half the height of a mature canopy tree. Yet elevating the surface only this much above the water table decisively alters both the species composition and the structure of the vegetation.

A seasonally flooded "igapo" forest (see page 25) grows on low-lying terrain adjacent to streams. The bases of the trees are typically underwater for several months a year during the rainy season. A belt of "igapo alto," or tall igapo, occurs on somewhat higher clays where better oxygen penetration into the substrate allows more active root metabolism.

Where the land rises above the seasonally flooded zone, the substrate can be either clayey or sandy. The clays, although quite infertile, are effective in holding water, whereas the sandy spodosols are very poor at holding either water or nutrients. Where the sand is shallow and overlies clays capable of retaining moisture through the dry season, the spindly, closely spaced trees attain moderate heights of 20 to 30 meters, creating a characteristic forest type known locally as "caatinga alta." Higher up, where the water table is more than 1 meter below the surface, there is a stunted version of the same formation, known as "caatinga baja," or low caatinga. Under the most extreme conditions of sandy knolls or ridgetops, the natural vegetation is spare, open, and hardly taller than a man. These stunted "banas" are a surprising anomaly in a climate that supplies more than 3500 millimeters (about 140 inches) of rain a year.

Other types of upland soils are found in the San Carlos region. Oxisols occur where the surface soil has a high content of oxidized reddish clays. These support a mixed "tierra firme" forest of high species diversity, but the trees do not attain more than moderate height because the roots have difficulty penetrating the dense

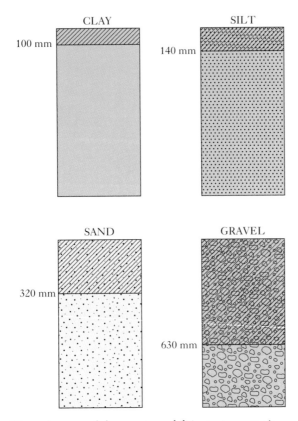

Fine and coarse soils have contrary abilities to store water in wet and dry environments. Where the rainfall is plentiful enough to saturate the soil, clays imbibe great quantities by swelling. Gravel soils, in contrast, can store only limited amounts of water in the spaces between grains. Therefore, in a dry spell in an otherwise wet climate, coarse soils dry out first. The opposite occurs in dry climates, where most of the scant rainfall comes in the form of brief showers that barely wet the surface. All the water is absorbed superficially by a clay soil, which will then lose it to evaporation as soon as the sun emerges. Very little lingers long enough to be captured by plants. On the other hand, rain that falls on gravel or rock will trickle downward into deep crevices. Protected from the sun's rays, water stored below the surface is available to plants to the depth indicated by the diagonal lines on the profiles. Acacia trees consequently grow on stony soils in parts of the Sahara that receive only 200 mm of rain annually, but they require 500 mm to grow on clay soils.

clay. Somewhat taller, though less diverse, legume-dominated forests occupy sandier ultisols nearby where trees are able to send roots to greater depths.

These seven markedly distinct forest types occur within a radius of a few kilometers of San Carlos and form a complex mosaic over the entire region. Local residents and the indigenous Yanomami Indians recognize these types and take advantage of their varying properties in organizing their own use of the landscape. Compared to the basins of silt-laden whitewater rivers elsewhere in Amazonia, the blackwater Rio Negro drainage is poor and starved for nutrients. The Yanomami inhabitants of this region live in widely scattered settlements at a population density of less than one person per square kilometer, practicing shifting cultivation on patches of alluvial substrate. In contrast, human population densities approach 200 per square kilometer elsewhere in the humid tropics, as in Rwanda or on the island of Java, where fertile volcanic soils support high agricultural productivity.

Anthropologists argue about the ultimate factors that limit Yanomami population density. Until recently, the proximate factor limiting the numbers of Yanomami has been warfare between clans. The anthropological debate has revolved around the question of whether the warfare was motivated by a desire to acquire multiple wives or by the need to control huge areas to ensure an adequate game supply in a protein-poor landscape. Although the answers are not all in, it is clear that the Yanomami live at extremely low densities, even for Amazonia. It is difficult to avoid the opinion that the sparse population is related to the fact that the

A group of Yanomami men discuss the fine points of an arrow. All over the world, the crush of "civilization" is pushing forest people such as these from their homelands.

Yanomami live on some of the poorest soils on earth.

Tropical Forests as Leaf Factories

Earlier, we saw that tropical forests are among the most productive ecosystems on earth, so one might reasonably suppose that they would be wood factories par excellence. Instead, tropical forests are excellent leaf factories; their wood production rates do not exceed those of other types of forests. This counterintuitive finding has a two-part explanation.

Tropical forests carry more leaves per unit area (higher leaf area indices) than other forests. The reason lies in their highly stratified structure: these forests are constructed of many vertically superimposed crowns. The crowns of all but the uppermost stratum of trees are partially or wholly shaded and hence capable of only weak photosynthesis. Productivity is concentrated in the upper canopy, where crowns are fully exposed to sunlight. Unlike their tropical counterparts, the trees of northern forests are mostly intolerant of shade, so many fewer individuals occupy positions below the canopy. The productivity of trees in a temperate forest is therefore more uniformly high, whereas in a tropical forest there is greater variation. More of the energy captured by a tropical forest goes into producing leaves, many of which, being in the shade, provide a rather meager return.

A second consideration is that tropical forests exist in a perpetually warm climate where

Respiratory Costs for a Tropical and a Temperate Forest

Forest	Location	Aboveground biomass (tons per hectare)		Losses from respiration (tons per hectare per year)		Maintenance costs (tons per ton per year)	
		Wood	Leaves	Wood	Leaves	Wood	Leaves
46-year-old beech forest	Denmark	129	2.7	4.5	4.6	0.035	1.7
Lowland dipterocarp forest	Pasoh, Malaysia	414	7.6	18.8	29.1	0.045	3.8

respiration rates are high. Plants respire by taking in oxygen and releasing carbon dioxide, just as animals do. The energy derived from respiration is used in transforming sugars into wood and other tissues. Some sugars are consumed simply in maintaining living tissue, and, in the high ambient temperatures of the tropics, it costs a plant more in respired carbohydrate to maintain a given complement of leaves and wood. Add to this the fact that many temperate forest trees are dormant for nearly half the year, and the difference in the respiratory cost of maintenance becomes quite significant. In the comparison illustrated in the table above, the tropical forest had three times the weight of wood and leaves of a temperate beech forest. The respiratory cost of supporting this extra mass, however, was four times greater for wood, and six times greater for leaves. Tropical forests are obliged to run faster in the heat, and consequently there is less profit to invest in wood production.

Farming in the Humid Tropics: Do the Benefits Justify the Costs?

All around the world governments are sponsoring programs for colonizing the tropical forest. Sometimes the motivation is to establish a national presence in remote frontier zones. But more often it is to satisfy the thirst for land generated by an ever-expanding population of peasant farmers. It is understandable that governments feel compelled to respond to the needs of the landless, but is clearing the rain forest truly the best available option? Can rain forest lands provide permanent gains in the ability to feed growing populations? If so, then perhaps the gains can offset the cost of an irrecoverable loss of biodiversity. If not, then the whole world is the poorer.

The science of ecosystem ecology makes it possible to apply objective criteria to the development process. If "development" results in the conversion of a biologically diverse and productive landscape into an impoverished and unproductive one, it seems unarguable that this is "bad" development. The criterion of "good" development has to be expressed in terms of sustainability. If future generations have the same intrinsic right to the earth's resources as our own, then there can be no ethical justification for engaging in nonsustainable activities.

Sustainability can be judged from the ability of the land to maintain a given level of productivity under the prevalent technology. The rules are simple and intuitive. Soil should not be lost to erosion faster than it is produced by weathering. Nutrients should not be lost to runoff or removed in the harvested product faster

than they are being restored through weathering and from the atmosphere. The use of fertilizer is a perfectly acceptable, and indeed often a necessary, means of ensuring sustainability.

We shall now examine two contrasting examples of "development" in the humid tropics to see how well they meet the criterion of sustainability. Our first example is the Lua' fallow system. The Lua' are a tribal people of the Philippines who have maintained themselves for centuries by practicing a traditional form of slash-and-burn agriculture. That the Lua' have persisted in the same region for so long is testimony to the sustainability of their agricultural practices. Virtually all the land is used, even that on steep slopes, but only for a single crop

Hundreds of millions of slash-and-burn cultivators live on the economic margins of civilization in the humid tropics, like the farmers who are burning this land in Peru. Many inhabit remote districts offering no access to schools, doctors, or transportation. A large majority of these people would prefer to live in cities if jobs were available.

of upland rice in each cycle of use. The key to sustainability is in the length of the fallow periods that separate each bout of cultivation.

The cycle begins with the selection of a patch of secondary forest for the year's effort. Clearing is initiated at the beginning of the dry season. The underbrush is cut first, and then the trees are felled, leaving stumps a half a meter to a meter tall. Many of these will later resprout to accelerate the regrowth of the forest. The fields are fired on a preselected date after the slash has dried in the sun for six to eight weeks.

The timing of the burn is important to the success of the crop. If too early a date is chosen, the larger trunks will not have adequately dried, and the burn will be poor. On the other hand, waiting too long entails a risk that an early thunderstorm will soak the slash. A hot burn ensures the maximum release of mineral-bearing ash from the fallen forest.

Rice is planted even before the rains begin. Despite the strong sun and elevated temperatures of the dry season, the soil is damp below 5 centimeters because there has been no forest on the land to withdraw the moisture. By taking advantage of the sunny period preceding the rainy season, the Lua' coax the maximum productivity out of their system. Arrival of the rains stimulates the germination of thousands of seeds stored in the soil, and weeds begin to compete with the rice crop. Shortly after the harvest, which occurs toward the end of the rainy season, the now fallow fields are a sea of herbaceous plants and tree seedlings. Meanwhile, many of the cut stumps have sprouted and sent up new leaders. The regrowth quickly takes up any nutrients remaining in the soil

and continues to benefit as the charred trunks of the erstwhile forest gradually rot away.

The removal of the one or two tons of rice per hectare that such plots produce, although seemingly trivial, so depletes the available nutrient stocks that a second planting is unrewarded. Full recovery is possible, but it requires a lengthy fallow. With the Lua', cycle times average 10 years. This means that at any given time, 90 percent of the terrain is lying fallow.

This scenario, or ones very similar to it, describes a traditional lifestyle practiced in the humid tropics around the world. The system is inherently sustainable, provided the fallow periods are adequate. So long as births and deaths remain in balance, sustainability is ensured. But with modern medicine increasingly available to populations such as the Lua', death rates have dropped, while birth rates have remained constant. As the population increases, the need to clear more land each year forces compromise. Fallow periods are reduced until declining harvests result in chronic malnutrition. In many countries an acute scarcity of land discourages the young from contemplating a future as traditional farmers and contributes to an accelerated migration to the cities.

Hungry, jobless urban masses generate an explosive discontent that frightens politicians, whether civilian or military. In scores of countries the response has been to build roads into the rain forest, where free land is made available to all comers. It is hoped that opening up new land will simultaneously alleviate unemployment and stimulate agricultural production. But at the same time, to appease city dwellers, governments provide massive subsidies of basic

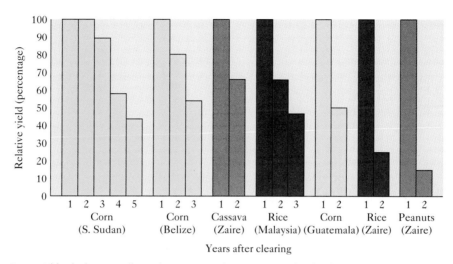

Crop yields decline rapidly with successive harvests on unfertilized tropical soils. Generally lacking fertilizer, tropical farmers are compelled to clear a new plot of forest every year to feed growing families.

food commodities, forcing retail prices below the local cost of production. Such contradictory policies lead to the double ills of chronic inflation and stifled incentive in the agricultural sector. Herein lies the seed of the Third World debt crisis.

Our second example illustrates development through the introduction of "modern" agriculture to the humid tropics. Experts in international development are acutely aware that the only way to generate the dramatic gains in food production needed to keep pace with population growth is to increase the productivity of the land. The slash-and-burn fallow system does not lend itself to intensification. That 90 percent of the land lies idle antagonizes the professionalism of First World agricultural experts, as does the inherent untidiness of the

system. A good field isn't full of stumps and charred logs. It is clear, and free of roots, stones, and other impediments to mechanization.

Well-intentioned aid missions have made valiant attempts to introduce Western technology to the humid tropics. In many instances the results have been disappointing. The reasons for lack of success have not always been clear, for seldom in such efforts have traditional practices been used as a control for evaluating the "modern" technology. One of the few studies using controls was an exemplary study in Nigeria. Instead of automatically assuming that Western methods would produce superior results, the investigators compared a graded series of methods that spanned the gap between traditional and modern practice.

On the all-important sustainability criteria, the traditional method won hands down. It resulted in less runoff, less soil erosion, and less nutrient loss than any of the mechanized modifications. But traditional practice does not include fertilization. Rapid declines in productiv-

Experimental fields at Yurimaguas, Peru, showing a kudzoo fallow in the foreground and upland rice in the background.

ity are therefore inevitable unless a long fallow period follows every cycle of cultivation. Mechanization, on the other hand, causes soil compaction and leads to unacceptable rates of erosion and nutrient loss. Are there other alternatives?

Professor Pedro Sanchez has been investigating alternatives at the North Carolina State University research station in Yurimaguas, in Amazonian Peru. He has pioneered two systems, termed high-input and low-input cropping. As the terms imply, the former is based on heavy applications of fertilizer, while the latter more closely approximates traditional methods.

Some tropical soils, while low in nutrient content, have physical properties that lend themselves to mechanization: good consistency, aeration, and percolation. High-input trials at Yurimaguas have demonstrated that such soils can sustain continuous cropping for at least several years. Some experimental fields have been producing three crops per year in a rotation of rice, peanuts or chick peas, and corn (maize) for up to 11 years. Although it is not yet clear whether the Yurimaguas system will prove a panacea for tropical agriculture, it clearly represents a major advance. If a field can be used for 5, 10, or 15 crops in succession, instead of just 1 or 2, before it has to be abandoned to fallow, the pressure on the remaining forest will be greatly alleviated.

Because it relies on heavy investments in machinery, fertilizers, and pesticides, the high-input system has been criticized as inappropriate to local needs. Peasant farmers have neither the training nor the capital to take advantage of its benefits. Moreover, transportation and the

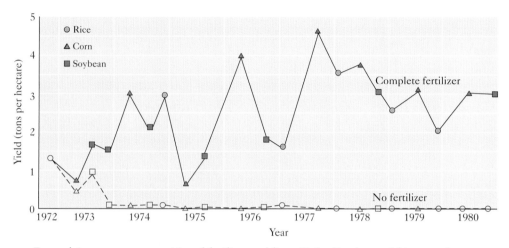

By applying generous quantities of fertilizer and lime, Pedro Sanchez and his research group were able to rotate rice, corn (maize), and beans in a cycle of three crops a year for nearly a decade without any discernible decline in yield. In the absence of fertilizer, productivity dropped to zero after the second year. Unfortunately, relatively few farmers in the tropics have enough free capital to purchase fertilizer, even where it is available.

prices offered by local markets are unpredictable, and the necessary supplies of fertilizers and other chemical agents are often unobtainable. Even when science has developed miracle varieties or improved technologies, the adoption of new ways is frequently resisted by highly traditional village agriculturists. What is needed, it was argued, is a more appropriate technology.

To answer these criticisms, Sanchez developed the low-input cropping system. Instead of relying on external sources of nutrients, every effort is made to retain the initial stocks. Cultivation is kept to a minimum for erosion control. All biomass that is not harvested is retained, including that of weeds and crop residues. Nutrient recovery is promoted by planting kudzoo between cropping cycles.

(Kudzoo is a fast-growing nitrogen-fixing vine in the legume family that has escaped widely into the southeastern United States.) The low-input method is labor intensive instead of capital intensive and thus meets most of the objections to the high-input system.

So far the results have been modestly encouraging. The productivity of some fields has been extended to seven cropping cycles over 3 years. The inevitable fertility declines do occur, requiring eventual retirement of the land to less intensive uses, but they are delayed. Even this is progress, for we are in a stalling game. Anything that can be done to prolong the period of intensive use of the land is a step in the right direction, for it means that less forest is cut in the meantime. Nevertheless, the ultimate goal of sustainability still seems a long way off.

The Global Diversity Gradient

During my childhood years as a fanatical collector of reptiles and amphibians, I was always struck that the northern destinations preferred by my parents for summer vacationing were nearly bereft of these creatures. Whereas as many as 120 species could be found in my native Virginia, all that Maine or Quebec had to offer were a few garter snakes, some frogs and an occasional painted turtle. All the really interesting reptiles and amphibians seemed to live in the south, just beyond my boyhood reach. A mere day's drive into North Carolina would have put me within range of a whole constellation of exotic beasts: sirens (eel-like aquatic amphibians), glass (legless) lizards, the Carolina anole (a lizard that changes color), soft-shelled turtles, crowned snakes (the

Scarlet (with yellow in the wings) and
red-and-green macaws jostle one another
at a salt lick in Peru.

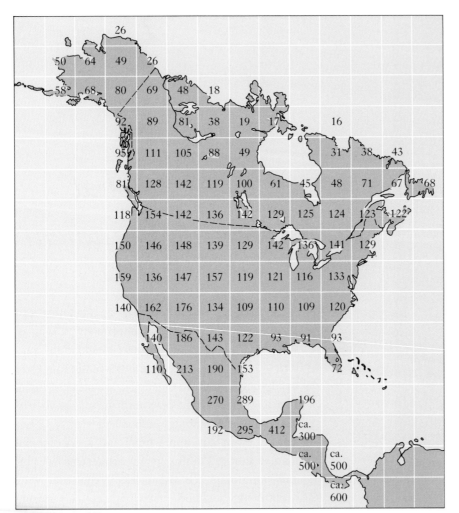

The numbers of bird species breeding in North and Central America in squares 350 miles on a side. Within the United States and Canada, the highest diversities occur in the mountain states of the West, where the varied topography creates a rich mosaic of habitats. Proceeding southward, the numbers begin to climb rapidly in central Mexico when tropical forests are encountered. Despite its tiny area, more land birds breed in Costa Rica than in the United States and Canada combined.

only rear-fanged snake to occur in the region), the American alligator, and many others. The farther south one went, the more there were.

The tendency for the numbers of species of organisms to increase toward the equator is by no means limited to reptiles and amphibians. It is true for trees, birds, and insects, to mention only a few examples, and for marine organisms as well. For many taxonomic groups the temperate-tropical diversity gradient (change in number of species with latitude) is steep, but for some it is rather shallow, and for a very few, such as salamanders and aphids, the gradient is reversed—that is, more species occur in temperate regions. Why, no one precisely knows, although the variety of patterns is our first hint that no single factor will suffice to "explain" tropical diversity.

The accompanying map shows that the regional density of bird species increases approximately fivefold from mid-latitude to the tropics. Much of the increase occurs precipitously upon entering the tropical belt at a latitude of 20 to 25 degrees. The "gradient" is thus not an even one, but appears more as a stepwise discontinuity that distinguishes the frost-free parts of the globe from those that experience seasonal cold.

Trees show a similar pattern. The richest temperate forests in North America, those of the southern Appalachians or the Gulf Coast, support at most 50 to 60 species, whereas any respectable tropical forest contains that many in a single hectare. With a few notable exceptions, data on larger areas in the tropics barely exist because of the years of effort required to discover and identify all the species.

Probably the best-known tropical flora is that of Barro Colorado Island, Panama, a 15-square-kilometer research reserve of the Smithsonian Institution. Investigators working on BCI, as the island is called, benefit immeasurably from a comprehensive botanical manual written by Thomas Croat of the Missouri Botanical Garden. A rarity for the tropics, this weighty tome describes some 1369 species of vascular plants found on BCI, including 365 species of trees. An area of similar size and topographic relief in Ohio, let us say, might have a tenth as many tree species. Moreover, the flora of BCI is not especially rich by tropical standards.

Trees manifest one of the greatest increases in diversity between midlatitude temperate and tropical regions, about tenfold on a spatial scale of a few hectares to a few square kilometers. On larger spatial scales, the difference is even greater. Birds, bats, and marine molluscs all increase roughly five times on the smaller spatial scale. In the light of such consistency, it is curious that quadrupedal mammals (four-legged species, to distinguish them from bats) increase by only a factor of two or so.

A colugo, commonly known by the misnomer of flying lemur, on a tree trunk in Sarawak, East Malaysia. It is not a lemur, nor does it fly. However, it does glide from trunk to trunk, aided by a supple web of skin that, when spread, stretches from hand to foot and foot to tail.

An olingo *(Bassaricyon gabbii)* searches for fruit and small prey in a Costa Rican cloud forest. The olingo, along with its relatives the kinkajou and coatimundi, are Neotropical members of the raccoon family.

Nearly all temperate mammals are terrestrial, rarely leaving the ground, or semiarboreal, spending part of their time on the ground and part in trees. True arboreal mammals, such as flying squirrels and porcupines, are in the minority. The increased diversity of quadrupedal mammals in the tropical forest is entirely attributable to larger numbers of arboreal forms, such as sloths, anteaters, kinkajous, and others in the New World; flying lemurs, civets, rats, and numerous squirrels in Asia; and civets, squirrels, anomalurids (resembling flying squirrels), and tree hyraxes, among others, in Africa. In addition to these, primates are the predominate group of arboreal mammals in all three regions.

Although the numbers of terrestrial birds, reptiles, and amphibians increase dramatically in the tropics, the numbers of terrestrial mammals do not. No one really knows why.

The Equitability/Stability Hypothesis

Biologists since the days of Wallace have attempted to understand the temperate-tropical diversity gradient. Many proposed explanations have focused on the tropical climate. A climate that was warm and free of either seasonal or longer-term perturbations would allow species to accumulate over eons. The north temperate region, in contrast, had not long ago suffered the trauma of continental glaciation, a fact that has been known to geologists for well over a century. Many species of large mammals, including the mammoths and mastodons, had died out during the late Pleistocene (or glacial epoch), and many specialists attributed the extinctions to climatic adversities. By this reasoning, the rapid and extreme climatic gyrations of the temperate zone would have the effect of periodically trimming diversity to lower levels, whereas the comparatively benign tropics seemed to ensure species survival.

Although this argument, or various elaborations of it, has an appealing ring, it also contains some hidden weaknesses. First, one must be careful not to confuse a seductive correlation with an explanation. Yes, the mild and equitable climate of the tropics does somehow seem more conducive to an exuberance of life than the capricious seasonality of the north, but in the absence of any deeper reasoning, this statement of the theory is merely a tautology. Second, steady improvements in the resolution of the fossil record began to cast doubt on the idea that the glacial advances per se could be held responsible for mass extinctions. For example, many of the plants that occupied central Europe before the onset of the Pleistocene can still be found in "refugia" on the southern slopes of the Pyrenees, Caucasus, and Elburz.

Using more accurately dated fossils, scientists confirmed that the mastodons and other now-extinct mammals survived the vicissitudes of the Pleistocene, dying out later when the glaciers were in full retreat. It is now widely acknowledged that the disappearance of large mammals is better attributed to overhunting by early man than to the stresses of a changing climate. The equitability/stability hypothesis, as it had been called, began to lose some of its luster. Not that it was entirely wrong. But the evidence for it seemed a lot less convincing, and, as stated, it offered only a tautological caricature of a theory.

Here is where matters stood in the early to mid-'60s, when ecologists began to consider the global diversity gradient in earnest. A blitz of new ideas appeared within a few years, but, as is often true in science, it took some time to sort them out.

Are There More Niches in the Tropics?

A major area of ecological theory is directed toward understanding the coexistence of species—how species with similar diets and behavior, such as the two crows in my backyard, can stably exploit the same habitat. If the requirements of two species overlap too much, then one will normally eliminate the other, at least where the two occur together. For two similar species to reside within the same habitat, they must partition the available resources in some way. If the species occupy adjacent but nonoverlapping space, they are not technically in coexistence and may pursue nearly identical lifestyles without jeopardizing one another's persistence in the ecosystem.

More than half a century ago, investigators coined the term "niche" to describe the requirements that must be satisfied by a species' environment. The niche is an abstract concept of primarily heuristic value. It allows us to imagine that every species plays a role in nature that is distinct from the roles of all other species. A niche is defined by the range of physical and biological conditions permitting a species' existence. We might imagine a certain type of mouse that could not tolerate excessive heat or cold, but that might nevertheless thrive in a seasonal climate by seeking shelter in a burrow when the temperature became intolerable. An adequate food supply is another requirement, but often a flexible one, in that many mice consume a broad diet of seeds, fruit, tender shoots, and insects. Tender shoots might be abundant in the spring, insects in the summer, fruit in the fall, and seeds in the winter, so the mouse would need all four to survive a yearly cycle. It may also require dense vegetation in which to evade predators. And if a superior competitor, or some other biological deterrent, such as a disease, occurred in some of the available habitats, the mouse might be excluded from these. Putting all these requirements together, our imaginary mouse species could live where the environment provided suitable substrate in which to dig a burrow, a continuous year-round food supply, concealing vegetation, and a habitat free of superior competitors and disease. Environments that satisfied all these overlaid requirements could either be widespread or highly localized, and the species would be distributed accordingly.

It must be noted that a particular niche is the property of an organism, not of its environment. A species possessing a broad niche might be able to occupy a wide range of environments, whereas one with a narrow niche would be more limited in some sense. Since a given niche consists of several attributes, it can be broad with respect to one, perhaps diet, and narrow with respect to another, perhaps the range of acceptable habitats.

The usefulness of the niche concept is that it enables investigators to formulate testable hypotheses. In particular, they can frame questions about the temperate-tropical diversity gradient in terms of niches. Do tropical habitats accommodate more niches, allowing greater numbers of species to coexist? Or, alternatively, do tropical species exhibit narrower niches than their temperate counterparts? This might be the case if they consumed more specialized diets or occupied narrower ranges of habitats. First posed in the early 1960s as a device for focusing research, these questions have led to significant advances.

Much of the subsequent progress has emerged in studies of tropical bird communities. Birds were the obvious organisms to study for a number of reasons. They strongly exhibit the temperate-tropical diversity gradient, they are observable, mostly diurnal in habits, and quantitative methods for studying them are well established. But most crucial, birds are one of the only groups of tropical organisms that can be identified reliably by nonspecialists in the field. A scientist working on any other group would have had to devote a major effort to collecting specimens. To this day, birds and mammals are the only major taxa for which tropical field guides are available. Tropical reptiles, amphibians, fish, plants, and of course insects cannot be identified with confidence except through direct comparison with specimens in a major natural history museum.

Tropical forests are structurally more complex than equivalent temperate habitats, and

A scythebill (genus *Campyloramphus*) searches for prey in an epiphytic bromeliad. Members of the woodcreeper family, scythebills also hunt in buttresses, bamboo, and cane, using their long bills to extract arthropods and small vertebrates such as frogs from their daytime hiding places.

their greater structural complexity may allow tropical habitats to accommodate more niches. Tropical forests are generally taller than temperate forests, contain a higher diversity of plants, and are richer in life forms, particularly lianas, palms, and epiphytes. It would be hard to imagine that all these components of habitat complexity did not contribute to the elevated numbers of bird species in tropical forests.

Structurally complex vegetation might accommodate large numbers of bird niches by permitting birds to seek food in more diversified ways. As they search for arthropods, birds perform many distinct types of operations on the vegetation. They pick at open surfaces, peck rotten wood, flake bark, poke into crevices, rummage in debris, glean foliage, and engage in any number of other activities in pursuit of prey. The bills, wings, feet, and other morphological characteristics of different bird species are adapted for the behaviors employed in these feeding activities. Some familiar examples are the chisel-like bills that woodpeckers employ for flaking and excavating, the long wings of swallows that confer the speed and maneuverability needed to capture insects on the wing, and the minimal feet of hummingbirds, minimal because they serve only for perching and not for walking or hopping.

If a wide range of food resources is available, birds may have similarly varied opportunities to specialize in what food they eat, as opposed to how they seek it. The diets of birds are remarkably varied. Different species emphasize fruit, seeds, nectar, or animal prey. The prey sought by birds are most frequently arthropods, but include snails, reptiles, amphibians, mammals, and other birds as well. The hunting and capture methods of birds are as diverse as the escape tactics of their prey. Different types of prey may fly, jump, creep, crawl, scurry, or freeze to avoid capture. Some birds are highly specialized in what they eat or how they find it, and others less so. The possibilities for diversifying avian niches thus seem almost unlimited.

Leaf-gleaning birds need sharp eyes to spot a leaf katydid (family *Tettigoniidae*), which relies on cryptic coloration to evade predators in a Peruvian forest. Katydids and their relatives in the insect order *Orthoptera* constitute the principal prey of many birds and several monkey species.

Just how avian niches are diversified, however, was by no means obvious in 1960. Choice of diet could provide a critical distinction. Alternatively, tactics for finding and capturing prey might prove more basic, or perhaps choice of habitat was the point of departure. It required the intuition of an experienced naturalist to make an educated guess.

The ecologist Robert MacArthur of the University of Pennsylvania made this guess. In his dissertation research he had shown that some North American wood warblers living in Maine spruce forests were extraordinarily sensitive to subtle nuances of habitat. One species searched for prey only at the spirelike tips of spruce crowns, another concentrated on the outer foliage of lower branches, while still another was more attracted to inner branches closer to trunks. These observations suggested that structural features of the habitat were basic to defining a bird's niche.

If MacArthur were correct, habitats that offered a wide array of foraging substrates would be able to accommodate the feeding strategies of a greater number of bird species than structurally simpler habitats. Having a canopy, an understory of shrubs, and a layer of herbaceous plants at ground level, a temperate forest offers foraging opportunities to pecking woodpeckers, trunk-creeping nuthatches, gleaning vireos, hawking flycatchers, and walking ovenbirds. By offering a greater array of micro-habitats, a forest could thus be expected to support more bird species than an open field.

In a series of now classic papers, MacArthur and his coworkers demonstrated a strong linear relationship between habitat complexity and bird diversity in the eastern United States. This result affirmed the insights derived from his observations on warblers, and it suggested to MacArthur a possible basis for understanding tropical bird diversity. If similar measurements of avian diversity and habitat complexity could be carried out in the tropics, the results would reveal whether tropical birds resembled temperate birds in their response to habitat. Perhaps the greater structural complexity of tropical forests would by itself account for their more diverse bird communities.

MacArthur's hypothesis was the first serious attempt to break out of the circularities inherent in the equitability/stability hypothesis. A decisive result would indicate, at least for birds, that tropical diversity could be explained by differences in the physical structure of the environment. Unfortunately, MacArthur's efforts to test this proposition in the 1960s foundered on technical difficulties. Over time, however, a picture has emerged that has clarified some of the principal issues.

Investigators have taken two approaches to the problem. One has been to determine whether structurally simple habitats in the tropics harbor the number of bird species predicted by the empirical relationship determined for North American habitats. Such simple habitats include tree plantations composed of only one or two species of plants and lacking all the exotic life forms of the primary forest. Despite the simplicity of these "habitats," they proved to contain more bird species than occur in any temperate habitat, no matter how complex. Clearly, the physical attributes of the habitat are not the whole story, but they could still be a part of it.

The second approach was to see whether MacArthur's relationship between habitat complexity and bird diversity, when extrapolated to values of habitat complexity found in mature tropical forests, would predict the diversity of the bird communities of these forests. Temperate forests, even the tallest and best preserved, harbor no more than 30 to 40 bird species. A doubling or tripling of this diversity might be predicted from the greater structural complexity of tropical forests, depending on technical assumptions employed in analyzing the data. Instead, recent research has documented that bird communities in Amazonian Peru may contain more than 200 species, an increase of five- to sixfold over the temperate level. Again, the greater structural complexity of tropical habitats appears to offer at best only a partial explanation of tropical bird diversity. What other factors might then be involved?

More and Larger Guilds

For some fresh ideas we must go back to MacArthur's intuitive deduction that avian niches are defined by the structure of the habitat. His insight successfully accounted for the different avian diversities of North American habitats, but faltered when extended to the tropics. We may recall, however, that there were other possible criteria for discriminating niches, diet being one, and the behavior performed in gathering food another. To explore these possibilities, we compare tropical and temperate bird communities by making use of the "guild" concept.

As the word implies, guilds are groups of species that share a common livelihood. The members of a guild may be united by their diet—for example, all the species in a community that feed on nectar. Or, alternatively, they may be grouped by behavior—all species that search for prey while clinging to vertical trunks, or all species that sally out to catch flying insects on the wing. Because neither diet nor behavior alone provides fine enough distinctions to be useful, guilds are commonly defined on the basis of diet and behavior jointly. When a community is classified into its component guilds, the effect is to isolate groups of species that resemble each other ecologically more than they resemble the members of other guilds. For this reason, guild members are often actual or potential competitors.

The value of the guild concept is that it allows us to compare communities composed of species that are taxonomically unrelated to one another. Because the members of a guild share a common habitat, diet, or behavioral trait, they can be regarded as filling a "guild niche."

Armed with these new concepts, we may now explore a second approach to the question of why more bird species are found in tropical than in temperate habitats. Tropical habitats may indeed proffer a larger number of ways for birds to make a living, much as a large city offers more diverse employment opportunities than a small town. To examine this possibility, we shall compare the guild organization of the bird communities inhabiting a temperate and a tropical forest. The two forests were carefully

A spider of the genus *Pandercetea*, an attractive prey for bark-gleaning birds, is nearly invisible on a variegated background of bark and lichens on the trunk of a tree in Malaysia.

chosen to be as similar as possible so that we eliminate the effects of differing structural complexity, accepting the fact that a perfect match is an impossibility. We shall ignore the effects of structural differences between the habitats at the start, and later consider where they might play a role.

The two bird communities to be compared occupy forests in Peru and the southeastern United States. The Peruvian site is representative of forests that cover much of Amazonia. The temperate forest merits some further comment. Located near Columbia, South Carolina, in the Congaree National Monument, it is one of the last virgin floodplain forests in North America, and by far the largest. The trees in the Congaree are larger on average than those in a typical tropical forest, some of them reaching heights of 45 meters and diameters of 2.5 meters. Although it does not contain as many tree species as a tropical forest, the Congaree floodplain supports more than 40 species, considerably more than most temperate forests. In structure, it more closely resembles its Amazonian counterparts than any other forest remaining in North America. Several strata of trees provide a complex vertical organization embellished with epiphytes (Spanish moss and resurrection fern) and heavy lianas (grapes and others). There is even a palm in the understory. Here is structural complexity to equal that of many tropical forests.

The virgin forest of the Congaree National Monument harbors a breeding bird community of 40 species, 43 if one counts the extinct Bachman's warbler, Carolina parakeet, and ivory-billed woodpecker. The extant species fall into 16 guilds, of which only 3, the woodpecker, flycatcher, and foliage-gleaning guilds (warblers and vireos), contain more than two species. In other words, the species are highly dispersed among guilds, suggesting that not many species are direct competitors in this community.

In the Amazonian bird community, species are even more equitably dispersed among guilds, but now all but 2 of the 24 guilds contain more than 2 species. The community total, 207 species, is five times that of the warm temperate locality, and many individual guilds contain multiples of the number of species in the corresponding temperate guilds.

Comparing the two communities, it is immediately evident that the tropical forest supports species in 8 guilds that are not represented in the temperate forest. These guilds contribute a total of 56 species, accounting for one-third of the "extra" bird species found at the tropical site. To what features of the tropical environment can the presence of these 56 species be attributed?

One species is a nightjar (similar to a whippoorwill) that eats moths, 8 are parrots that feed on the seeds of immature fruits still developing in the trees, 3 are terrestrial frugivores (fruit eaters) that scavenge ripe fruit after it has fallen to the ground, and the remaining 44 belong to several insectivorous and omnivorous guilds. Why should none of these guilds be represented in the temperate forest? The question compels us to engage in post-hoc explanations, a practice that is scientifically frowned upon because it tempts rationalization. As the circumstances leave no alternative, we shall proceed, hoping to exercise due caution in arriving at conclusions.

Evergreen tropical forests produce many kinds of food resources that flourish all year round, including many that are consumed by birds, such as fruits, seeds, nectar, and of course arthropods. Temperate forests produce all these resources, but only on a seasonal schedule. The availability of each type of resource consequently gyrates through cycles of boom and bust. Fruits are most plentiful in summer and fall, seeds in fall and winter, nectar in spring and summer, and arthropods only in the warm months. To survive in an environment where the feeding opportunities are constantly shifting, birds of the temperate forest must either switch seasonally from one type of

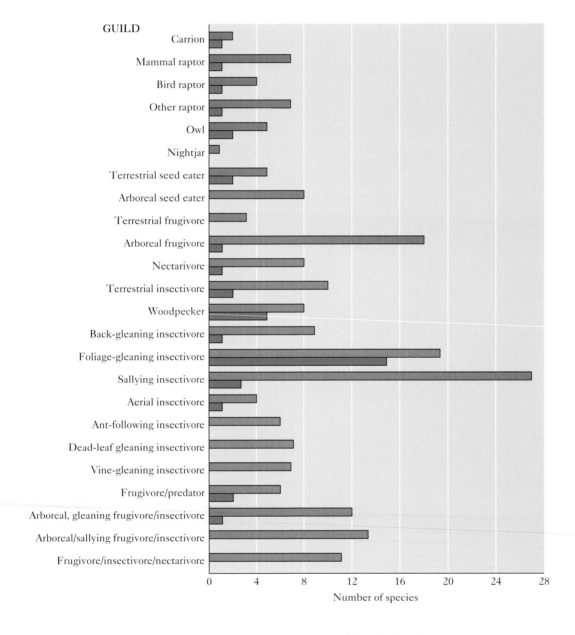

The bird communities of a temperate and a tropical forest broken down into guilds—groups of species that have similar diets and behavior. The tropical community *(red bars)* contains more guilds and more species per guild than the temperate community *(blue bars)*.

resource to another, or they must migrate. Some do both. The familiar American robin that arrives in New England in the spring to hunt earthworms on suburban lawns spends the winter feeding on holly berries in southeastern forests. Without such versatility, life in the temperate zone would be almost impossible.

In the tropics, however, species can afford to specialize on a narrow class of resources, and by doing so they are perhaps able to compete more successfully for their chosen food. The bill of a parrot, for example, is good for cracking seeds, but not for capturing insects. A hummingbird can reach nectar in the deepest recesses of a flower, but would be at a loss to separate the seeds from the pulp of a fruit.

Other tropical birds, members of the omnivore guilds, are versatile, but not in the fashion of temperate birds. Many tanagers, honeycreepers, and other species of the Neotropical forest consume a mixed diet of insects, fruit or nectar, or all three. When nesting, the adults of many of these species visit fruiting or flowering trees to supply their own metabolic needs quickly, leaving time free to search for protein-rich insects required by their rapidly growing nestlings. This behavior is relatively uncommon among birds of the temperate forest, presumably because they nest during late spring and early summer, when insects are most abundant, but before most kinds of fruit have ripened. Again, because its seasons are less distinct, the tropical forest offers opportunities for dietary specialization that are not available to residents of the temperate forest.

This argument does not explain the absence from the temperate community of three guilds of insectivores that are well represented in the tropical community: 6 species that snatch

A resplendent quetzal *(Pharomachrus mocinno)* swallows a fruit of the laurel (avocado) family in the cloud forest at Monteverde, Costa Rica. Specialized frugivores such as the quetzal abound in tropical forests, thanks to a year-round supply of fruit.

fleeing insects ahead of the onslaught of army ant swarms; 7 species that search curled-up dead leaves that have lodged in vines and branches; and 7 species that live in dense vine tangles. The greater seasonality of the temperate environment again provides the answer, but in a somewhat different form.

Army ants *(Eciton burchellii)* carrying a wasp larva in the tropical forest of Mexico.

Hundreds of thousands of army ants may form a single swarm that spreads through the leaf litter, amoebalike, flushing out startled arthropods. Cockroaches, spiders, centipedes, crickets, and many other frightened prey hop or scurry over the dead leaves in their attempts to escape. As they are driven into the open to flee, these normally hidden denizens of the litter become vulnerable to capture by birds waiting at the leading edge of the advancing swarm. Whereas swarm-raiding army ants are common in the Neotropics, they do not occur in the southeastern United States, perhaps because arthropod prey are lacking during the winter. Birds that specialize in searching hung-up dead leaves ("arboreal leaf litter") would also find themselves without a source of food through most of the year because dead leaves do not begin to accumulate until the end of the season and quickly blow out of the trees after

winter arrives. By a similar argument, birds that hunted principally in vine tangles would find meager foraging opportunities in temperate forests compared to many tropical forests, where vines tie together large sections of the canopy.

Virtually all the "extra" guilds of the tropical forest bird community are tied to food resources or foraging substrates that are scarce or absent for long periods in the markedly seasonal temperate environment. Seasonality doubtless accounts for some near-empty temperate guilds as well. A seasonal lack of fruit almost certainly explains why only one frugivore inhabits the temperate forest, in contrast to the 18 present at the Amazonian site. Similarly, the ruby-throated hummingbird is the only hummingbird to occur in the Congaree, whereas 8 species occupy the Peruvian forest. Again, contrasting levels of resource availability appear decisive. In the Congaree floodplain the ruby-throated hummingbird will find only one significant source of nectar during the nesting period, a vine known as trumpet creeper. Other plants bearing suitable flowers are scarce or absent. In the Amazonian forest, however, there are scores of plant species that attract hummingbirds, some with long tubular flowers accessible only to long-billed species, others with short flowers catering to short-billed species. Suitable flowers are borne by trees and vines in the high canopy, and by treelets and herbaceous plants in the understory. Nectar, in short, is available in many forms from top to bottom of the multitiered forest, offering an array of possibilities for specialization on the part of nectar-feeding birds.

Fewer than 40 percent of the birds of the Amazonian forest feed primarily on plant products, and the proportion is even lower in the temperate forest. In both forests, insectivorous birds predominate. Invariably, the tropical guilds contain more species than their temperate counterparts. Here, seasonality does not seem especially relevant because insects reach peak abundance in temperate forests during the avian breeding season. Insect diversity is lower in temperate forests, perhaps as a consequence of reduced plant diversity, but the numbers of individuals may be high. Insects are therefore no less available to insectivorous birds in temperate forests than in tropical forests.

What factors might then account for the larger numbers of species in the tropical guilds? There are at least two possibilities: first, the tropical forest offers larger guild niches, and second, the tropical guilds are more tightly packed with species. The first of these calls for an ecological explanation, the second for an evolutionary explanation.

The bill of a scintillant hummingbird (*Selasphorus scintilla*), although relatively short, is long enough to drain the nectar from the shallow flowers of this Costa Rican epiphytic heath (family Ericaceae).

Are Tropical Guild Niches Larger?

The members of a guild typically consume items of food that span a range of qualities or sizes. Consider, for example, how species in the guild of frugivores might partition the fruit supply. A small frugivore can swallow only small fruits, whereas a large frugivore can consume fruit of almost any size, and so it enjoys a broader range of choice. Yet the large

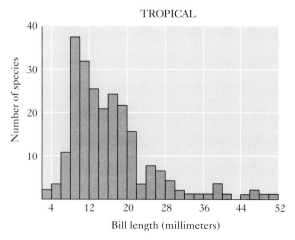

TROPICAL

Number of species

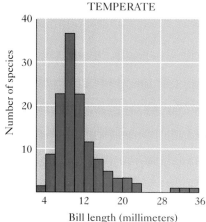

TEMPERATE

Number of species

Bill length (millimeters)

Bill lengths of birds breeding at latitudes of 8 to 10 degrees north *(left)* and 42 to 44 degrees north *(right)*. On average, the tropical birds have longer bills than do their temperate counterparts.

frugivore faces limitations of a different kind. Consider yourself, for example. If you were hungry and could visit a blueberry patch or a peach orchard, which would you choose? Unless you happen to dislike peaches, the answer is obvious. You could fill up on peaches in 15 to 20 minutes, but picking the same quantity of blueberries would take at least an hour. It is true that bears do frequent blueberry patches, but more generally large frugivores pass up small fruits in favor of large ones. Thus guild members often "partition" resources on the basis of size, thereby avoiding excessive competition.

The question of how differences in body size mitigate competition between species is an important one theoretically. The seminal contribution to this topic was made by the eminent Yale ecologist G. Evelyn Hutchinson. While studying copepods (barely visible crustaceans known as water fleas) in some lakes in northern Italy, Hutchinson noticed that lakes contained varying numbers of species in varying combinations, but that the community of any given lake was composed of species that differed in size. With striking consistency, each species was 1.2 to 1.3 times as long as the next smaller.

Such size differences are now recognized in ecology as conforming to the "Hutchinsonian

ratio." Hutchinson surmised that a ratio of approximately 1.25 in linear measurements represents the barely sufficient size difference that allows two closely related species to coexist and subsist on a common pool of resources. Hutchinson's idea is now often referred to as the "law of limiting similarity." In a broader context, it conveys the notion that two species cannot coexist unless they differ by some minimum degree. A Holy Grail of theoretical ecology has been to derive the Hutchinson ratio from first principles.

Hutchinson's conjecture implies that there is a limit to the number of species that can subsist on a given pool of resources. If the resource pool is diverse, a greater diversity of consumers will be able to share that pool. As an illustration, we may consider the guild of arboreal seed eaters. In the Amazonian forest, this guild is occupied by a spectrum of parrots, ranging from a tiny sparrow-sized parrotlet, through somewhat larger parakeets, to full-sized parrots, and ending with the colorful and majestic macaws. Because they are of different sizes, these birds are able to partition a resource they all exploit in common: seeds, particularly immature seeds. The diminutive parakeets seek the tiny seeds of figs, for example. The birds adroitly manipulate the fruits in their incredibly versatile bills, patiently extracting the seeds and discarding the pulp.

While this activity might be perfectly rewarding for a parakeet, it would be disdained by a parrot or macaw. A larger bird would not be able to manipulate the tiny seeds with the necessary precision, nor would the bird derive sufficient energy to satisfy its metabolic needs even if it could. Larger members of the guild are attracted to larger fruits, many of which are protected against just such depredations by hard or fibrous exteriors. A macaw, for example, has no difficulty opening a Brazil nut, but in nature it rarely has the opportunity to eat one. Brazil nuts come packed in a heavy-walled, coconutlike shell that is impervious even to so formidable a device as the bill of a macaw. Nevertheless, macaws are able to open a wide array of protected fruits to obtain the seeds within. By making use of its larger, stouter bill, each parrot species in the size progression gains access to resources that are beyond the capability of the next smaller species. Size thus serves as the most important basis for resource partitioning within many guilds.

In a given environment, the range of fruit sizes or the hardness of their protective coverings would define the breadth of the available resource pool. Any number of consumer species from zero to many could be dependent on a given resource pool, but in general a broader resource pool would be expected to support a greater number of consumers.

The existence of larger guild niches was long ago proposed as a partial explanation of tropical bird diversity by Thomas Schoener, now at the University of California, Davis. The possibility came to light in a comparative study of the bill sizes of tropical and temperate insec-

A blue-crowned motmot (*Momotus momota*) prepares to swallow an *Anolis* lizard. The motmots are one of several families of large-billed birds that prey on large arthropods and small vertebrates in tropical forests of the New World.

tivorous birds. For birds of a given size, the tropical species consistently possessed larger bills. In addition, there were more large species with very large bills.

This finding suggested that tropical insects might on average be larger than temperate insects. With his colleague Daniel Janzen, Schoener sampled the insect communities of some forests in Costa Rica and the United States. The tropical samples turned out to have many more large insects, including species that were larger than any in the temperate samples. In light of this, it seemed reasonable to conclude that these outsize insects were providing niches for large-billed insectivorous birds in the tropics. Support for this conclusion is found in the observation that the largest species in each of the Amazonian insectivorous guilds exceeds the body weight of its temperate counterpart (except in the bark guild, where our pileated woodpecker is slightly larger). In general, however, the differences are rather small, implying that tropical guild niches are only fractionally larger than their temperate equivalents.

The Packing of Tropical Guilds

A much more conspicuous difference between the temperate and tropical guilds is that the latter routinely contain twice to several times as many species. In the case of insectivorous guilds, it is hard to attribute the "extra" species to the effects of seasonality, to greater tropical tree diversity, or to larger guild niches. For each guild represented in the temperate forest, there simply seem to be more species in the tropics.

The presence of larger numbers of species in the tropical guilds brings up one of the questions that launched this discussion, the possibility that the niches of tropical birds are narrower. But to conclude that the niches of tropical birds are narrower simply because there are more species would be tautologous, and therefore incorrect. This is why it is important

to make the comparison by means of the guilds, because guilds are defined independently of the species within them. Since several of the tropical guilds contain many more species than their temperate equivalents, the niches of the component species do appear to be narrower. In the jargon of ecology, species are more tightly "packed" into the tropical guild niches.

The most plausible way for tighter packing to arise is through speciation, the evolutionary mechanism of proliferation of species. As we shall find in Chapter 6, evolution appears to have generated more species in the tropics, and as a consequence many of them have become narrowly specialized.

The towering height and internal structural complexity of the tropical forest provide the setting for finer specialization. An excellent example is found in the Neotropical antwrens, a group of small foliage-gleaning birds, analogous to our temperate warblers. Up to 10 species can occupy a single locality. Very similar in morphology, and often in superficial appearance as well, antwrens feed on insects but search for them in a variety of different ways. At the Amazonian site used for our tropical/temperate comparison, one species lives near the ground, another in the canopy, and several more in the middle tiers of the forest. One of the latter forages in aerial leaf litter, another inspects the bare surfaces of twigs and branches, and three more hunt prey in dense foliage.

The illustration on the facing page shows how four separate species that glean live leaves segregate into narrow and only partially overlapping foraging zones. Each seeks prey in a place or in a way that differentiates it from all the others. Such fine partitioning of the structural

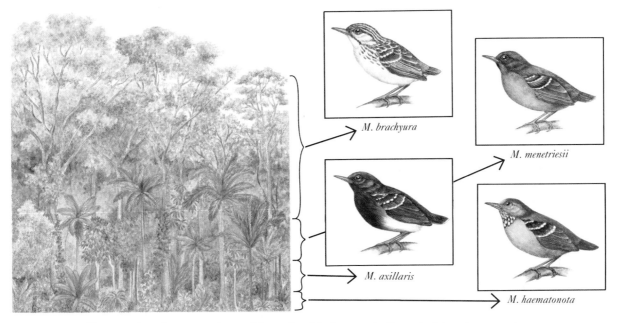

Four species of antwrens (genus *Myrmotherula*) show vertically stacked foraging zones in an Amazonian forest. Such close ecological relationships are an indication of tighter species packing in tropical guilds.

environment might not be possible in a more simply constructed habitat. Here is where the greater structural complexity of the tropical forest seems to come into play in accommodating diversity. By choosing slightly different heights at which to forage, the four antwrens avoid competition sufficiently to coexist in close proximity. Such finely tuned ecological roles are typical of species living in diverse tropical communities.

At each step of the way through this chapter we have found answers in the details of natural history, what a species eats, how it finds food, how it relates to other species in the same community. A quest that began with some disarmingly simple questions has led us into a thicket of complexity. Every suspected factor, upon inspection, seemed to hold part of the answer to the larger question.

Although there is much yet to be learned, it is clear that many factors contribute to the "extra" diversity of the tropics. These include a more complex structural habitat (illustrated by the four antwrens and the guild of specialized vine foragers), additional guilds made possible by the less-seasonal environment (arboreal seed eaters, terrestrial frugivores, followers of army ants), greater tropical plant diversity (providing resources for hummingbirds and frugivores), larger insects (allowing broader guild niches for insectivorous birds), and finally, in the close packing of many guilds, evidence of exuberant speciation. The sources of diversity in groups other than birds can be expected to be similarly complex, although the details are likely to vary.

If, in retrospect, the initial questions—Do tropical habitats offer more niches? Are the niches of tropical species narrower?—now seem naive, it is a measure of the progress that MacArthur inspired. Naive beginnings require no apology, since the simplest questions are sometimes the most penetrating.

4

A Mosaic of Trees

In our efforts of the previous chapter to understand the animal diversity of tropical forests, we kept coming back to the plants—the complex structure of the habitat and the great variety of edible plant products in the form of leaves, flowers, fruits, and seeds. If it were not for the plant diversity of the tropical forest, it is safe to say the animal diversity would not be nearly so great. We cannot fully understand tropical diversity without appreciating the factors, evolutionary and ecological, that contribute to plant diversity and its perpetuation through time.

The approaches we employed in discussing animal diversity cannot be applied in the same form to plants. Animals subsist on many types of food resources and seek them in distinct places and ways. In contrast, all green plants

Many tree species respond to the dry season by losing their leaves, flowering, and sprouting new leaves, as in this forest in Rondonia, Brazil.

require the same essentials for growth: sunlight, carbon dioxide, water, and mineral elements. Moreover, plants occupy fixed positions in space. The adaptive features that differentiate the ecological roles of plant species are therefore very different from those that allow animals to coexist in complex communities.

Carbon dioxide, the basic ingredient of photosynthesis (roughly analogous to the food of animals), is everywhere in the atmosphere. Because the atmosphere is in constant flux, one plant cannot significantly reduce the amount of carbon dioxide available to another; competition for this basic resource is therefore nonexistent.

Competition for sunlight is largely a matter of first come, first served, because sunlight is intercepted out of a directional stream. It is rare to see two tree crowns intermingling, because the one that is higher can generally outgrow any potential competitor beneath it. Vines, however, are quite capable of stifling the growth of trees by overgrowing their crowns and smothering the foliage. Like sunlight, water must also be intercepted from a directional stream. Competition for water, however, is not simply a matter of precedence, because water does not pass by at the speed of light. It penetrates the soil and must be absorbed by roots. A plant that invests heavily in roots can be a good competitor for water, but, in doing so, it inevitably becomes a poorer competitor for light, because it has less material to invest in developing a stem and crown. A parallel argument applies to competition for nutrients in the soil. Recall from Chapter 2 the expanded root masses of trees growing on nutrient-poor spodosols in Venezuela.

Because light, water, and nutrients are required by all plants, it is hard to imagine that

partitioning of these resources accounts for more than a part of tropical plant diversity. One need note only that some of the most diverse plant communities on earth occur in per humid climates where the soil is permanently saturated with water. Diversity is not conspicuously depressed, in other words, when the possibility of competing for water is eliminated. Similarly, there is no clear tendency for plant diversity to be higher on poor soils, where competition for nutrients must be intense, than on rich soils, where most important nutrients are available in abundance. Without denying that plants compete for light, water, and nutrients, it seems clear that to coexist in communities, plants must partition other features of the environment. In this chapter I shall explore a number of approaches in an effort to identify these additional features.

I shall begin by examining patterns of tropical plant diversity on some different spatial scales—intercontinental, regional, and local. At the largest scale, that of entire continents, differences in diversity are likely to reflect the accumulated results of evolution. The three major tropical regions of the earth have been largely isolated from one another for the past 80 to 100 million years, and that long space of time has given those regions ample opportunity to evolve distinctive species. Then, at a regional level within continents, we shall see that patterns of plant diversity correspond to the principal components of environmental variation, such as rainfall and soil chemistry, but not to any one component in particular. Finally, at the smallest scale, that of nearest neighbors in a stand of trees, biological processes begin to play important roles in perpetuating plant diversity.

Intercontinental Patterns

The tropical portions of Asia, Africa, and the Americas form three great biogeographical units that differ markedly in size, geological history, geographical organization, and topographic complexity. Because there are so many differences, a priori there is no means for deducing which region might support the highest diversity of tree species. Ever since the first comprehensive analyses of contiguous plots were published several decades ago, botanists have widely held the opinion that the world's most diverse forests were found in peninsular Malaysia and northwestern Borneo. Single hectares in Sarawak and Indonesian Borneo could boast as many as 200 species, whereas the richest sites then known in Amazonia could claim only 175. Africa came in a weak third, a fact that seemed reasonable in view of the smaller total area of forest and a history of drought and fragmentation.

The conventional wisdom that the forests of Southeast Asia were the most diverse in the world has recently been challenged by Alwyn Gentry, a virtuoso botanist of the Missouri Botanical Garden. Gentry has contributed massively to the available information on the floristic richness of New World forests through his exploration of numerous sites in Central America, Colombia, Venezuela, and, especially, western Amazonia, a region that had not featured in previous botanical work, but which was known by ornithologists to harbor the world's richest bird communities.

Gentry has discovered what may prove to be the world's epicenter of botanical diversity near Iquitos, Peru. Hectares situated on both

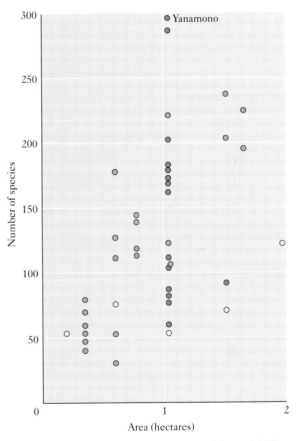

Tree species diversity in tropical forests around the world. The points represent the numbers of tree species with a minimum diameter at breast height of 10 centimeters found in plots of various size. Green points: Southeast Asia; red points: Neotropics; yellow points: Africa.

low- and high-fertility soils contained tree diversities of 289 and 300 species respectively, 30 percent higher than any reported from Southeast Asia. The high-fertility plot at Yanamono on the lower Rio Napo is truly incredible. The 300 species were contained among a total of only 606 stems exceeding 10 centimeters in diameter at breast height. Every other individual, in other words, represented a new species. Forty-eight species were counted in the first 50 individuals sampled. It is hard to imagine higher diversity than this!

The task of obtaining data such as these is by no means a simple one. An experienced botanist could go through a hectare of forest in Wisconsin, for example, and in one day identify every tree and still have enough time left over for an afternoon round of golf. Not so in the tropical forest. Most of the trees cannot be reliably identified to species from the ground, even by an expert. To be certain of the identifications, it is necessary to obtain voucher specimens of nearly every tree. This requires a dedication so unshakable that the researcher is willing to put his life in jeopardy, for the specimens can be obtained only by climbing the trees—hundreds of them. When Gentry analyzed a hectare at my field site in Peru, it took him nearly a month. During this period he climbed hundreds of trees, a feat of mental and physical exertion that left him exhausted. No wonder there are not many completely analyzed hectares of tropical forest, and that few plots of greater than 1 hectare have ever been sampled!

Even when specimens are in hand for each of the 600 to 800 trees that typically occupy a hectare, the job is far from done. The specimens must be taken back to a world-class herbarium and then sorted, first to family and then to genus. The specimens are then packaged and mailed out, genus by genus, or family by family, to perhaps the only human being competent to identify them. Likely as not, the specialist is burdened with dozens of such requests, each of which must await its turn. It might be a year or two before the specimens

come back with names sanctioned by the expert. Specimens lacking flowers or fruits often cannot be identified at all. These formidable physical and logistical bottlenecks in the analysis of tree plots have seriously held back our knowledge of tropical forests.

From a statistical standpoint, the proper response to high species diversity would be to increase the size of the areas sampled. A sample of 1 hectare fails to include many of the species present in the richer communities. A better sampling unit would be 10 hectares, yet the daunting task of obtaining voucher specimens from 6000 or more trees has all but precluded work at this scale. Instead, samples of a tenth of a hectare are more typical. When the available samples contain only a quarter of the diversity of the community being studied, it is literally hard to see the forest for the trees.

Although little enough is known about the trees of most tropical regions, even less is known about the other plant forms. These include herbs, vines, epiphytes, parasites, and treelets, diminutive understory trees too small to meet the minimum criterion used in most tree plots of 10 centimeters diameter at breast height. In most tropical forests, these other plant forms contribute more than three quarters of the total plant diversity. At the Rio Palenque reserve in Ecuador, a site receiving 3000 millimeters of rainfall annually, trees make up only 15 percent of the flora. The most diverse groups at Rio Palenque are herbs and shrubs (36 percent of the species) and epiphytes (22 percent). Herbs and epiphytes are so prominent because the perpetually moist conditions allow plants to cling to exposed surfaces such as trunks and limbs without desiccating.

Our inquiry into tropical plant diversity shall be focused on trees even though trees are only a minority component of the total plant diversity of tropical forests generally. The limitation is forced on us by the almost total lack of information on other plant forms.

Regional Patterns

There exist only two credible data sets on the composition of tropical forests on the scale of an entire region, one for northwestern Borneo and the other for western Amazonia. The former is the more comprehensive and is accompanied by extensive soil analyses, so we shall consider it first. These two perspectives lead to conclusions that differ on some major points, but converge in support of the notion that forest composition is regulated by identifiable features of the environment, particularly soil chemistry and rainfall.

The first data set pertains to the Malaysian state of Sarawak and the Sultanate of Brunei, both located in the northwestern quarter of the island of Borneo. Peter Ashton of Harvard University and several colleagues are currently analyzing inventories of over eight hundred 0.2-hectare plots. Nearly all the plots were established more than twenty years ago by the Sarawak and Brunei Forest Departments, not in the interest of advancing basic scientific knowledge, but to inventory the timber resources of the region. Unfortunately, despite the public outcry over tropical deforestation, nowhere in the world today is there any comparable effort being made to study tropical forests.

For their analysis of species richness, Ashton and his associates used 205 plots sited at 16 localities. The localities were selected to represent the full range of soil and climatic conditions available within the region sampled. The plots combined form a total area of 103 hectares and include 75,000 individual trees having a circumference at breast height of at least 12 inches. Altogether, the plots contain an incredible 3200 species. Ashton and his colleagues estimate that this number represents about 80 percent of the tree species known from the lowlands of northwestern Borneo.

Although the plots were individually small, the large number of them provided numerous replicates (repeated samples) at each locality. Soils ranged from nutrient-rich clays derived from basalt (a volcanic rock), to white sand spodosols, to acidic peat swamps. A statistical procedure known as "association analysis" compared the plots on the basis of the similarity of their tree species compositions. Replicate plots tended to cluster closely together in the analysis, as did relatively distant plots representing similar soil conditions. In short, given information about the characteristics of a site (especially soil type), the presence or absence of particular tree species becomes quite predictable.

What do Ashton's results tell us about species diversity? When Ashton analyzed samples of 1000 trees from localities scattered over the whole region, he found that the numbers of species in a sample varied from about 120 to more than 350. None of this variation in species richness was correlated with climatic factors—not surprisingly, since all the sites are located at low elevation and receive ample rainfall. Instead, soil chemistry appears to be the

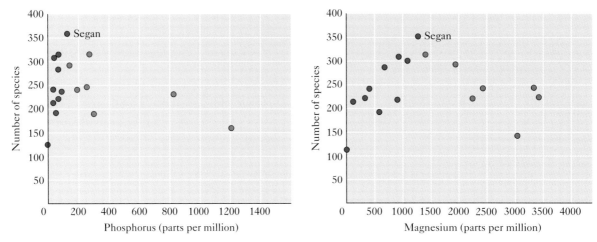

Tree species diversity in Sarawak, East Malaysia varies with the level of phosphorus and magnesium in the soil.

main determinant of tree diversity within the region.

To investigate the influence of soil chemistry, species richness was plotted against the concentration of macronutrients in the soil. The most species were found at sites having intermediate nutrient levels. Soils at the Segan Forest Reserve, the site boasting the highest diversity, were intermediate in their content of phosphorus, magnesium, and potassium. On poorer soils, species richness tended to increase with nutrient levels, whereas on soils richer than those of the Segan Forest Reserve species richness declined as nutrient levels rose. Species richness varied more from site to site on nutrient-poor soils than on nutrient-rich soils, suggesting high sensitivity to nutrient levels at the low end of the scale, and more finely differentiated niches.

These are powerful and interesting results that must be taken into account by any theory of tropical plant diversity. The interpretation

offered by Ashton and his colleagues is roughly as follows. Growth on poor soils is limited by the availability of nutrients, and only species adapted to nutrient scarcity can reproduce on such sites. Competition for light is consequently diminished because trees invest more in extracting nutrients from the soil. Ashton acknowledges the likelihood that, on the poorest soils, some portion of the diversity of the 1000 tree samples is due to small-scale variation in nutrient levels within the sample plots. In contrast, on soils of high fertility growth is limited more by access to light than by nutrient availability. Species with high growth rates are able to monopolize the canopy, reducing the light available to smaller species and thereby lowering diversity. On soils of intermediate fertility, plants compete for both nutrients and light, leading to maximum diversity.

The interpretation suggested by Ashton and his colleagues is ecological in that it rests on interactions between species that take place in a contemporary framework of time. Yet one can easily frame an evolutionary argument to explain the same pattern. Suppose that the environment of East Malaysia encompasses a certain range of soil conditions. Over millions of years, evolution will have generated species that are best adapted to different conditions within the existing range of variation. If sites possessing extreme soil properties are rare, relatively few species will evolve to occupy them. More species will be able to compete successfully on common soil types offering conditions intermediate between the extremes. With the information at hand, it is not possible to distinguish between these two alternative interpretations.

The second regional data set we shall consider was compiled by Alwyn Gentry for western Amazonia. Because it is more nearly the work of a single individual, the perspective it provides is quite different. At most sites, Gentry was obliged to lay out and analyze the plots by himself with few or no assistants and without the support of a soil analysis laboratory.

For this reason, it is not possible to analyze his results in a manner parallel to that followed by Ashton and his colleagues. This lamentable inconsistency is not to be attributed to a lackadaisical attitude toward procedure on Gentry's part, but to the meager levels of support this type of work is able to attract from granting agencies, often barely sufficient to pay airfare.

Where Ashton and his associates found consistency and predictability of species composition, Gentry found very little. Six 0.1-hectare samples from the Iquitos region in upper Amazonian Peru all possessed very high species diversity (163 to 249 species), yet, other than the two replicate plots at Yanomono, the samples shared few species in common. Similarly, at the opposite end of Peru in the Tambopata Nature Reserve, 1-hectare samples located on different soil types had largely distinct complements of species, whereas a pair of replicate plots shared about half their species. Gentry concluded that the extraordinary number of woody plant species to be found in Amazonia is largely attribut-

Number of Species Shared by 1-Hectare Forest Samples near Iquitos

	Yanamono no. 1	Yanamono no. 2	Yanamono tahuampa	Mishana lowland	Mishana campinarana	Mishana tahuampa
Yanamono						
Terra firme no. 1	212	91	20	24	12	14
Terra firme no. 2		230	20–21	19	9	8
Whitewater tahuampa			163	9	5	ca. 19
Mishana						
Lowland noninundated				249	55	17
Campinarana (white sand)					196	3
Blackwater tahuampa						168

able to the presence of many distinct habitats within this vast region. He further suggests that dramatic differences in composition, but not species richness, are the consequences of specialization for different soil conditions.

Although Gentry contrasted his results with those of Ashton, I am struck more by their similarity. Ashton found that species were specialized for narrow ranges of soil conditions, and so did Gentry. Had Gentry been able to exploit the statistical resolving power afforded by a sample of hundreds of plots, including numerous replicates, he and Ashton might have arrived at similar conclusions.

Gentry's results from across the Neotropical realm reveal a strong and previously unsuspected pattern of great theoretical significance. Whereas Ashton found no relationship between species richness and either total rainfall or rainfall seasonality, Gentry found a strong and statistically robust relationship. Diversity in 0.1-

hectare plots increases linearly with rainfall at sites receiving from 500 mm to about 5000 mm (20 to 200 inches), whereupon the relationship levels off. At the upper end of the scale is a plot at Tutunendo, Colombia, that receives 9000 mm (350 inches!) of rainfall annually.

Here again the perspective of the two authors differs. Malaysia and Borneo experience the most equitable climate on earth, and rainfall within the region varies little. All the sites studied by Ashton receive between 2800 and 4800 mm of rainfall annually (100 to 190 inches), a range of only 1.7-fold. All Ashton's localities lie at the upper end of the scale studied by Gentry, which encompasses a variation of nearly 20-fold. That Ashton found no correlation of species richness with rainfall does not mean that Asian vegetation responds differently to climate, only that the variability needed to expose the relationship is not available in East Malaysia.

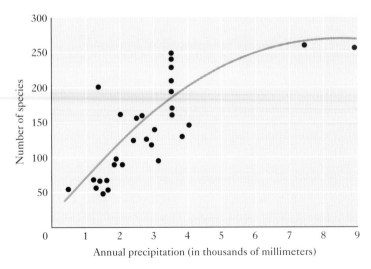

The number of species of trees in 0.1-hectare plots of Neotropical lowland forest plotted against annual precipitation. Only trees measuring at least 2.5 centimeters in diameter at breast height were counted. Small midstory and understory tree species contribute most of the increased diversity in the wetter sites; canopy diversity remains nearly constant throughout.

Gentry's result is important theoretically, for it undoubtedly holds one of the keys to understanding the high plant diversity of the humid tropics. What it means, however, is open to debate. It is not even clear that rainfall, per se, is the relevant variable, because in the Neotropics there is a strong inverse correlation between seasonality and rainfall. In other words, the length of the annual dry season is closely predicted by total rainfall. The greater the rainfall, the shorter the dry season, a rule of thumb that holds up to climates receiving 3000 to 4000 mm of total precipitation (120 to 160 inches). This is just about where Gentry's curve levels off, and where climates cease to include regular dry periods, suggesting that the occurrence of an annual dry season may seriously limit plant diversity.

The Scale of Nearest Neighbors

We come now to an examination of pattern at the scale of individual plants. Here the view is very different from the one we have had before, and the focus is on specific mechanisms rather than on broad correlations.

The simplest hypothesis one could make in reference to any particular site of uniform quality, given its climate, soil chemistry, water-holding capacity, and so forth, is that there should be one species among all those available that would prove the best competitor. Over time, that species should be able to monopolize more and more of the available space and eventually should come to dominate the site. (Each species plays the role of extreme specialist, as

in the example of the two islands in Chapter 1.) Other sites nearby might differ in slope, nutrient content of the soil, or in other qualities of importance, and these sites would be dominated by different species. The landscape at equilibrium would thus be a mosaic of single-species stands, each occupying sites offering a particular set of conditions. Such monotypic stands indeed characterize the coniferous forests of mountains in the western United States.

This simple model seems absurdly irrelevant to a forest containing 300 species of trees per hectare. Nevertheless, it provides a useful point of departure for further speculation. That the model clearly does not apply to species-rich tropical forests suggests that one or more of its underlying assumptions is invalid. By examining these assumptions it is possible to generate a set of hypotheses that can later be tested with the available evidence.

One questionable assumption is that the physical environment is unchanging over the entire surface of a sample plot. No site is truly uniform. Underground features such as nutrient availability, soil texture, acidity, drainage, and depth to water table might vary on small scales, such that even the most carefully selected plot would include a mosaic of conditions favoring different species at points only a few meters apart. Let us call this the microspatial heterogeneity hypothesis.

Another possibility is that the environment varies not in space but in time. No climate is absolutely monotonous. Each tiny parcel of ground in which a seed might germinate could experience varying conditions of soil moisture, humidity, or illumination. The seedlings of certain species might become established in the

This California redwood grove is typical of many western conifer forests in that it is dominated by a single tree species. The rare single-dominant forests that occur in the tropics are usually associated with anomalous soil conditions.

dry season, while those of other species might be at an advantage in the wet season. Or, the same idea could be rephrased in terms of dry years and wet years. We shall call this the temporal heterogeneity hypothesis.

Another assumption of our simple model is that tree species compete in such a way that over time one will predictably dominate any particular site. This assumption could be vio-

lated in two ways. First, different species might not compete. For this to be true, the species in a given stand must be equally able to replace themselves in the next generation. Each would have an equal chance of increasing or decreasing, as in a "zero sum game." This is the "non-equilibrium" hypothesis favored by Stephen Hubbell of Princeton University. In the absence of any competitive advantage, the populations of all species will fluctuate randomly without constraints, and the species composition at a given site is expected to vary unpredictably over both time and space.

As a second possibility, the species in a stand might compete in an orderly way, but there might not be enough time to achieve the equilibrium. The intervals between generations of trees are long, often longer than those of human beings, so competitive processes that take place over several to many tree generations would be difficult to measure. Moreover, major disturbances such as droughts, fires, and wind storms, although rare, might nevertheless occur often enough to disrupt the competitive process. Occasional catastrophes could maintain a stand in a condition of perpetual disequilibrium by preventing a dominant species from increasing to the point that other species were excluded. This is a somewhat modified statement of the "intermediate disturbance" hypothesis proposed a number of years ago by Joseph Connell of the University of California at Santa Barbara. This is also a "non-equilibrium" hypothesis in its assumption that stands rarely, if ever, achieve a composition that solely reflects the outcome of competitive interactions. It differs significantly from the Hubbell hypothesis in assuming that predictable changes in composition will occur between major disturbances.

Even within a physically uniform space, it is possible that a varied environment is created by biological processes. Trees may die in different ways, for example, and be replaced by species with distinct "regeneration strategies." A tree that dies standing up leaves a relatively small gap in the canopy, while one that is blown down, and perhaps carries several others along with it, opens a major hole in the forest. A shade-tolerant species is likely to succeed in the first case; a species that can take the most advantage of open sun to grow very rapidly will succeed in the second case.

Finally, different individuals of the same species can exert influences on one another's growth, survivorship, and reproductive success, resulting in additional forms of spatial heterogeneity. Seeds may survive poorly in the shadow of the parent plant, for example, and the growth of saplings may be suppressed in the vicinity of an adult tree of the same species. We shall refer to such interactions under the term "distance dependence," because their strength varies with the distance between individuals. Distance-dependent effects tend to intensify as a species increases in abundance and to diminish as it becomes rare. By this means, distance dependence can exert a strong stabilizing influence on the community. A species that becomes common will have more difficulty reproducing, but as its population declines these difficulties will disappear. A given population may thus fluctuate around some equilibrium density. The population densities at which distance-dependent forces will cause declines in the next generation may vary widely from one species to another. The mechanism of distance dependence can thus be expected to generate a broad range of abundances within a community.

If all this is too technical, the main point is that no species can be too common, and that rare species have a greater likelihood than common ones of increasing in the next generation. If no species can become common enough to monopolize the available space, then competitive interactions are relaxed, and additional spe-

A hopeful seedling unfolds its first leaves on the rain forest floor. At this stage, seedlings are vulnerable to herbivores such as deer, rodents, and pigs, and many fail to survive even one year. Those that do may then remain in suspended animation for many more years awaiting a light-giving gap to open in the canopy overhead.

cies can participate in the community. The nature and intensity of the distance-dependent forces that operate in a particular environment could serve to regulate the species composition in a predictable fashion. This is the antithesis of Hubbell's non-equilibrium hypothesis, so it shall appropriately be termed the equilibrium hypothesis.

By examining the assumptions undergirding the simple model described above, we have been able to formulate a series of hypotheses about factors that might regulate the tree species diversity of tropical forests. Our next task is to evaluate the hypotheses with empirical evidence on the organization and dynamics of tropical tree communities.

Dispersion of Tree Populations in a Costa Rican Dry Forest

Perhaps the most elementary question one can ask is whether members of the same species aggregate or avoid one another in space. Nearly everyone who has taken an elementary course in plant ecology will have gone through this exercise. One samples a plot of forest and then determines whether small subsamples contain two or more individuals of a given species more or less often than would be expected by chance. Another approach is to ask whether the nearest neighbors of randomly selected individuals themselves randomly represent the surrounding forest. If individuals of the same species do turn out to be randomly spaced in relation to one another, the result is not particularly enlightening. Strict randomness, however,

is not a likely result, as it is only one outcome in an infinite spectrum of possibilities. If individuals are more evenly spaced than random (hyperdispersed in the manner of an orchard), it suggests that a repulsion mechanism is at work. If, to the contrary, individuals are in close proximity (clumped) more often than would be expected by chance, the interpretation is more equivocal. Either the terrain within the plot is heterogeneous, and species are selecting particular microsites within the mosaic, or history has played a role in the sense that young trees tend to accumulate around particularly successful matriarchs.

Although this simple question had been answered for countless temperate forests by generations of college undergraduates, not until a decade ago, when ecologists began to exploit high-speed computers, was it answered for any tropical forest. A widespread opinion at the time was that the tree populations of tropical forests were hyperdispersed, because many species were so rare. The impression may have come down from Alfred Russel Wallace, who eloquently expressed a familiar reaction to the Malaysian forest.

> If the traveller notices a particular species and wishes to find more like it, he may often turn his eyes in vain in every direction. Trees of varied forms, dimensions, and colours are around him, but he rarely sees any one of them repeated. Time after time he goes toward a tree which looks like the one he seeks, but a closer examination proves it to be distinct. He may at length, perhaps, meet with a second specimen half a mile off, or may fail altogether, til on another occasion he stumbles on one by accident.

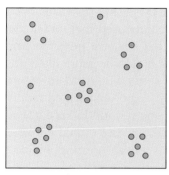 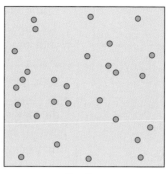 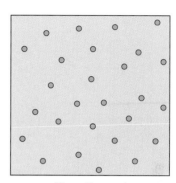

Clumped   Random   Hyperdispersed

The individuals in a tree population can be dispersed in clumped, random, or hyper-dispersed distributions.

This impression was bolstered by early at-tempts to inventory plots in Amazonia. It was found that more than half the species were present at population densities of less than one individual per hectare. At such low densities, a plot of 1 hectare, or even several hectares, would not provide large enough samples of most species to show how they were distributed in space. A more massive effort would be re-quired, something on a scale that had not pre-viously been attempted.

Stephen Hubbell took up the challenge. Selecting a well-studied tract of dry forest in Costa Rica's Guanacaste Province, he and his associates mapped every individual plant with a stem diameter of at least 2 centimeters in a 420-by-320-meter plot (13.44 hectares). They recorded a total of 135 species of woody plants, including 87 trees, 38 shrubs, and 10 vines. This is a low diversity compared to that of most tropical evergreen forests, but a much higher one than is found in any temperate forest.

Dispelling once and for all any lingering impressions derived from less intensive efforts, Hubbell's results showed that none of 61 spe-cies tested was hyperdispersed. Forty-four spe-cies exhibited significant clumping of adults, while the remaining 17 species had patterns that could not be distinguished from random.

Hubbell's results seemed to contradict a hypothesis that had been proposed by Daniel Janzen of the University of Pennsylvania. Jan-zen, along with others, had observed that seeds were less likely to survive in the proximity of parent trees than when dispersed to more dis-tant locations (a distance-dependent effect). Seed survival near the parent tree could be ex-tremely low because the fallen seed crop tended to attract large numbers of "seed predators." These range from small boring insects (Bruchid beetles and weevils, for example) to large verte-brates (squirrels, agoutis, peccaries). Janzen's hypothesis supposed that seed predators would search diligently under the parent tree where seeds were abundant, but would invest dimin-

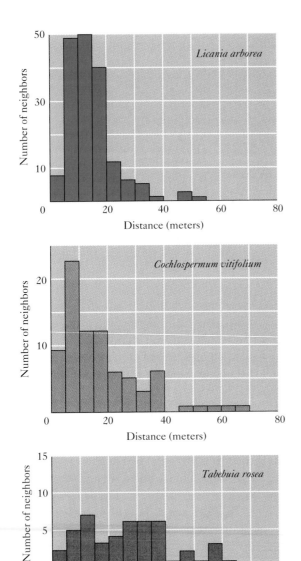

The distribution of distances from each individual tree of the indicated species to its nearest neighbor of the same species in a 13.44-hectare plot of dry forest in Guanacaste, Costa Rica. The three examples illustrate differing levels of clumping, from greatest to least: *Licania arborea*, *Cochlospermum vitifolium*, *Tabebuia rosea*.

ishing efforts at increasing distances as rewards declined. Janzen proposed that under these circumstances, as the distance from the parent tree increased, seeds would be more likely to survive, but fewer seeds would fall to the ground per unit area. At some unspecified radius from the parent tree, the decline in seed abundance away from the parent tree would be offset by increasing seed survival, creating a zone where the density of surviving seeds would be at its highest. If it is assumed that the surviving seeds germinate to produce seedlings, Janzen's model imagines a ring of maximal seedling density surrounding the parent tree. The distance of such an optimal zone of seedling recruitment from the parent tree would vary from one species to another, depending on how widely its seeds were scattered and on how its particular suite of seed predators behaved. Although Janzen did not explicitly make the claim, suppressed recruitment near adult trees could result in a hyperdispersed population.

Hubbell's data failed to uncover even a single example of hyperdispersion—in nearly every case he found juveniles to be more common near adults than farther away. Do these findings rule out Janzen's seed predation hypothesis? Not necessarily, because the hypothesis explicitly states that there is a distance from any parent tree at which seedlings are most likely to occur. It is important to note that the implications of this mechanism are not the same for common and rare tree species.

When a species is sufficiently common, Janzen's mechanism should result in fewer young trees growing near adults because the zones around parent trees that are intensively searched by seed predators will fuse, and fewer

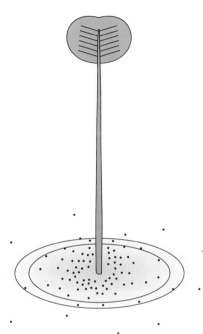

(Left) The ring around the parent tree represents the region where the density of surviving seeds is highest. Within this ring, the greatest establishment of seedlings is expected to occur. *(Right)* Fruit of a Neotropical nutmeg (*Virola* sp.), showing the large seed encased in a bright red aril that attracts the toucans and other birds that are its principal dispersers. *Virola* seeds are lost to seed predators in much higher proportions near the parent tree than farther away.

seeds will survive over the entire area of high adult density. But at the opposite end of the abundance scale, a different result is expected. In a rare species, the radius of the ring where the probability of seedling establishment is highest will be less than the mean distance between adult nearest neighbors. Thus for rare species the Janzen hypothesis does not predict hyperdispersion; it predicts clumped distributions as seedlings sprout closer to the parent than to other members of the population.

In fact, this is precisely what Hubbell found: the rarer the species, the more the individuals tended to be aggregated in space. With the exception of seven rare "outlier" species that were not reproducing on the plot, the inverse relationship between clumping and abundance was extremely strong and consistent. Hubbell's results thus provide vigorous support for Janzen's hypothesis at the low end of the abundance scale.

What happens at the high end? To answer this question, Hubbell examined the relationship between the number of juveniles per adult and the number of adults per hectare for the 30 most common species in the plot. In all 30 cases, there were fewer juveniles per adult in hectares containing larger numbers of adults than in hectares with smaller numbers of adults. The trends were statistically significant in 17 of these cases. Again, the prediction derived from Janzen's hypothesis was upheld.

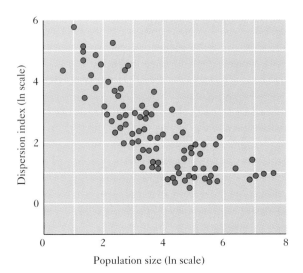

Trees in a Costa Rican dry forest are less likely to be clumped together as population size increases. Rare species are strongly clumped, whereas the individuals of common species tend to be randomly arrayed in space (dispersion index of 1).

We cannot conclude directly from Hubbell's results that Janzen's hypothesis alone accounts for the dispersion patterns of all tree species in the Costa Rican dry forest. Nevertheless, Hubbell's work lends it strong support as a regulatory mechanism of potential importance. It contains the essential ingredient of distance dependence needed to establish and perpetuate an equilibrium state: diminished reproductive success at high levels of abundance and enhanced success at low levels. It is, moreover, a mechanism that can operate over a wide range of abundances and that can potentially account for the chronically low densities of many species. The latter point is particularly crucial, because another puzzle in the tropical diversity problem is why there are so many rare species.

Although we are far from arriving at a final answer to this basic question, it is gratifying to sense that there has been encouraging progress toward understanding some of the fundamental underlying mechanisms. Seed predation is not the only important biological process to operate through an inverse distance relationship. Carol Augspurger of the University of Illinois has elegantly demonstrated that host-specific fungal pathogens can produce high mortality in seedlings near the parent plant and that isolated seedlings are more likely to escape attack. In addition, there is now mounting evidence that seedlings growing close to the parent tree are more vulnerable to herbivores than their more isolated siblings.

In principle, all such distance-dependent mechanisms should lead to the stabilization of a community's composition. It is not necessary that distance-dependent seed predation, pathogen-induced seedling mortality, and herbivory act in concert, or that they all act on every tree species. What is required is that one or another of these processes operate on most tropical tree populations, especially on the most abundant. For any given species, it is likely that one mechanism will be more powerful in limiting densities than others, although the relative strengths of these processes may vary from year to year. More seeds might escape predation in years of high seed production, for example, while we would expect seedlings to be more severely attacked by fungi in particularly wet years. Most critical is that these mechanisms prevent any species from becoming too common. Rarely does the most abundant species in a tropical forest occupy more than 10 to 15 percent of a stand.

Small forest ungulates such as this greater mouse deer (*Tragulus napu*) in Sabah, East Malaysia, often function as both seed predators and seed dispersers. Some seeds are chewed and digested, while others pass intact through the animal. Often germination is hastened by exposure to animal guts, and seedlings may profit by germinating in a heap of fertilizer.

The weight of evidence reviewed thus far seems to support the view that tropical tree communities move toward an equilibrium, but some of the hypotheses stated above have yet to be evaluated. These include diversification of regeneration strategies, and microspatial and temporal variability, none of which are simple phenomena. We shall next consider the issue of spatial variability.

Gap Dynamics

Even the most uniform sample plots contain obvious spatial heterogeneity both above and below the ground. Above ground, illumination may vary widely from spot to spot, depending on how recently branches or entire trees have fallen and created holes through which direct sunlight can penetrate into the understory. Below ground, there can be variability in soil depth, texture, and chemistry that is hidden from the human eye. Subtle differences can be extremely important. For example, almost imperceptible variations in topography on low-lying ground can decisively alter plant composition, as can the proximity of bedrock to the surface.

In relation to the question of tropical diversity, however, all these types of variability are relatively uninteresting, because there is no reason to think they are any greater in the tropics. They have been studied by plant ecologists in the temperate zone and are well known to affect the composition of plant communities.

The puzzle that confronts us is how, within even the most uniform plots of tropical forest, 9 out of 10 neighboring stems can be of a different species. Are there other kinds of microspatial variation that go beyond these obvious ones and that might be particularly relevant to the tropical environment? In fact, there do seem to be several, all brought to light by recent research. They pertain to what are called "gap dynamics" and "regeneration strategies."

We tend to think of trees as permanent fixtures in the landscape, living entities possessed of almost eternal vitality. The huge trees that inspire awe in some tropical forests enhance this impression. For those who live in the forest, the impression can be quite different. At my research station in the Peruvian rain forest, seldom does a week go by that I am not jolted awake by the startling sound, terrifying when close, of roots snapping under tension. The popping soon swells to a crescendo of creaks and groans that stirs primordial emotions. A cacophony of cracks and snaps accompanies a prolonged whoosh, culminating in a resounding thump, as a giant trunk slams into the ground, issuing a wave of reverberations that can be felt hundreds of meters away. At such moments I am thankful the sounds are no closer.

Despite the frequency of such occurrences, the tree mortality in this forest, at 1.7 percent per year, is close to average. At this rate, roughly 10 trees dies annually in each hectare. Some die standing up and gradually come apart, aided by the tireless exertions of termites. Others uproot, most commonly in wind storms or during the rainy season when the waterlogged soil offers little resistance. The rest simply break in two, frequently under the onslaught of a falling giant. The ensuing destruction opens gaps in the canopy, and it is in these gaps that most regeneration occurs.

Gaps can be small, as when a limb breaks out of the canopy, or impressively large, as when a vine-draped giant pulls and pushes a dozen other trees along with it in its fall, opening an area half the size of a football field. Most gaps, however, are small. The increased light stimulates seeds to germinate and initiates a race between new seedlings and saplings already present around the margins. If the gap is small, less than 10 to 20 meters across, trees around its edges will quickly extend their branches to fill the space, depriving freshly germinated seedlings of a chance to develop.

The appearance of a large gap, however, sets off a much more complicated set of reactions. Hundreds of seedlings begin to grow almost simultaneously. So-called pioneer species, which can increase in height up to several meters a year, almost invariably win the race, or at least the first lap. But because such trees rarely cast a deep shade or attain great heights, more shade tolerant species are able to grow up under them and eventually take their place. The advent of a large gap thus initiates a process known as patch dynamics. An even-aged cohort of trees quickly grows up to fill the space. The pioneer species eventually die out and are replaced by late successional species in a slow process that may continue for many years. On a larger scale, the forest as a whole is a mosaic of these microsuccessional patches of varying age.

Although gaps regularly occur in temperate forests, setting off similar microsuccessions,

The death of a giant can mean life for a host of smaller trees, as seedlings compete to usurp the sunlight available in a large tree fall gap in Peru. The decomposing twigs and branches of the fallen crown release scarce minerals, closing the nutrient cycle and enhancing the vigor of the regrowth. When loggers cut and remove trees, the nutrient cycle is broken, and the site remains impoverished.

more species participate in tropical microsuccessions. Accumulating evidence suggests that tropical trees may be more finely graded in their light requirements and growth rates. After the formation of a gap, several species may occupy the space before the ascendance of a long-lived canopy dominant finally returns a measure of stability to the spot. Another factor in the diversity of tropical microsuccessions is that tropical forests are able to accommodate more individual trees whose crowns overlap a given point on the ground. A microsuccession thus simultaneously includes species that will take their places as adults at different vertical levels. Unless one is keenly aware of the role each species plays in the three-dimensional structure, the presence of so many seems inexplicable.

If all this were not enough, several additional mechanisms have been discovered within the last decade or so that expand our understanding of how patch dynamics contributes to diversity. One finding is that large tree fall gaps can go through a period of "arrested succession," when the normal regrowth of trees is suppressed by a vigorous release of competing vegetation. Often the fall of a large tree can bring down festoons of lianas. The trauma is normally fatal to the tree but a windfall to the lianas, which can now spread over the mass of

A wand of bamboo (*Chusquea* sp.) in the Monteverde cloud forest of Costa Rica. Bamboos typically grow in treefall gaps, where they sometimes form dense thickets that can suppress tree regeneration for many years until the bamboos flower and die.

crushed branches and invade the well-illuminated borders of the opening. A recent gap can become an impenetrable mass of vinestems within a few weeks, stifling tree growth. The only trees likely to survive are certain species that periodically rid themselves of clinging vines by shedding their bark or lower branches. In my corner of Amazonia, bamboo aggressively invades tree fall openings with similar effect.

A flurry of interest in "gap dynamics" has revealed that openings of a given size are not all equivalent, and that gaps routinely contain internal heterogeneities that strongly select for different species in the regrowth. A gap that is aligned on a north-south axis, for example, receives less light than one having an east-west orientation. The direction a tree falls can thus influence the identity of the species that re-

places it. Similarly, the partly shaded margins of gaps favor a different set of species from those that do best in the fully illuminated centers. The most sun demanding "heliophilic" species are able to reach maturity only in the middle of large gaps. Because large gaps are relatively rare, it is no surprise that these extreme heliophiles are also rare in most tropical forests.

The demise of a large tree creates distinct zones in the resulting gap, as has been elegantly demonstrated by Aldo Brandani and his colleagues Gary Hartshorn and Gordon Orians at the University of Washington. Working at the La Selva Biological Station in Costa Rica, they have studied the pattern of regeneration in three zones: (1) around the disturbed soil thrown up by the overturned root mass, (2) in the narrow portion of the gap paralleling the fallen trunk, and (3) in the jumble of twigs and branches left by the erstwhile crown. In careful studies of 51 gaps, they found that different suites of species sprung up in the three zones, and that root zones accommodated fewer species than trunk or crown zones. In the tree fall gaps of temperate forests, tree species also tend to regenerate in distinct zones, but given the greater number of competing species the effects of this mechanism are likely to be much stronger in the tropics.

Regeneration Strategies

Foresters have long recognized a categorical distinction between tree species that regenerate, at least to the sapling stage, in the shade of a canopy, and those that have an absolute requirement for full sun during early establishment. Both temperate and tropical foresters have made this distinction, and both have imagined that most tree species could be clearly assigned to one category or the other.

This conventional wisdom has recently been overturned by Stephen Hubbell. After completing his research on the 13-hectare dry forest plot in Costa Rica, he took on the unprecedented challenge of mapping 50 hectares of mature tropical forest on Barro Colorado Island (BCI) in Panama. This prodigious undertaking brought tropical biology into the megabuck funding range for the first time in history, and the results have richly rewarded the investment. Each plant with a stem diameter of at least 1 centimeter in a 500-by-1000-meter plot was marked with a numbered tag, mapped, and measured. This effort required the participation of more than 150 students, volunteers, and hired staff over a three-year period. The project, which continues to this writing, has followed the recruitment, growth, and mortality of nearly a quarter of a million plants, representing over 350 species. The crucial and exacting job of supervising the identifications was performed by Robin Foster of the Smithsonian's Tropical Research Institute, the organization that administers BCI.

If indeed there were any doubters, the Hubbell-Foster 50-hectare plot has demonstrated unequivocally the necessity of studying tropical diversity on a large scale. As we saw earlier, Ashton's massive data set on the composition of East Malaysian forests was the first (and still the only) to reveal consistent species associations over a large region. Now, for the first time, the BCI plot has provided the statis-tical resolving power needed to study the reproductive and establishment biology of individual species of tropical forest trees. Even with more than 300 species in the plot, the average size of a sample for one species is nearly 800 individuals. Notwithstanding the large size of the average sample, there are still 25 species represented by a single individual in the total of 238,000 plants satisfying the minimum diameter requirement of 1 centimeter.

To investigate the light conditions under which different species become established, Hubbell and his associates obtained an indirect measure of light penetration into the forest. They began by estimating the height of the canopy over the intersections of grid lines spaced 5 meters apart within the plot, 20,000 points in all. For each 5-by-5-meter subplot, they obtained a single value for canopy height by averaging the measurements made at the four corners. Since each individual tree was mapped to one of the 20,000 grid squares, the average canopy height over all juvenile trees could be determined for the populations of each species.

The values calculated for canopy height ranged from less than 1 meter in fresh tree fall openings to well over 30 meters under the crowns of large trees. The juveniles of species that regenerate only in full sun were found primarily in grid squares with low canopies, whereas the juveniles of some other species were found principally in the shade of a high canopy. The species showing these two patterns of distribution correspond to the "sun" and "shade" regeneration "guilds" long known to foresters. With the acute resolving power afforded by large sample sizes, Hubbell and Fos-

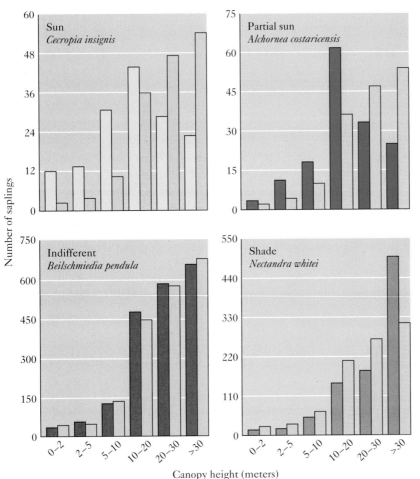

Representatives of four "regeneration guilds" in the tree community of Barro Colorado Island. In each figure the colored bars represent the distribution of canopy heights over 20,000 grid intersection points in a 50-hectare plot. Low canopies *(to the left)* are found only in recent treefall gaps, whereas high canopies *(to the right)* occur over much of the forest, wherever there are tall trees. The gray bars represent the distribution of saplings with respect to canopy height. Four types of distribution can be distinguished: sun, partial sun, indifferent, and shade.

ter were able to discriminate two additional regeneration guilds, hitherto unrecognized by foresters. Surprisingly, a majority of the tree species on BCI (171 out of 239) proved to belong to the two new guilds. The best-represented guild was composed of trees whose saplings grew in random locations with respect to the height of the canopy. Such species were said to belong to an "indifferent" guild. Finally, species whose saplings occurred principally under canopies of intermediate height were assigned to a "partial sun" guild. These results suggest that the traditional "sun" and "shade" guilds represent rather narrow specializations. In contrast, the species filling the "indifferent" regeneration guild can be considered generalists.

It is intriguing to imagine that the regeneration biology of trees in tropical forests is more complex than in temperate forests. If this were true, it might help to account for the higher species diversity of tropical forests. Yet no temperate forest has been studied with such powerful methodology, so it would be premature to conclude that tropical forests support additional "regeneration guilds." Although this seems likely, a definitive answer awaits the investment of an equivalent effort in a temperate forest.

We have seen that spatial heterogeneity can assume many forms. Recruitment patterns vary between large and small gaps, between the centers and edges of gaps, between the root, trunk, and crown zones of tree falls, between gaps and shaded locations in the forest interior, and between sites nearer and farther from an adult of the same species. Collectively these forms of spatial variability are certain to have a major influence on how the species composition of a forest is passed on from one generation of trees to the next.

The Time Factor

Just as the availability of sunlight, moisture, and other factors important to seedling establishment can vary over short distances, so can these same factors vary over time. In many regions, a regular alternation of wet and dry periods is an annual feature of the climate. Irregular events of less frequent occurrence, such as fires and floods, might also play major roles in the regeneration of certain species. We might thus anticipate that a variety of temporal patterns will contribute to diversity by favoring the recruitment of different sets of species.

It is conceivable that seeds may be programmed to germinate during a particular season to take advantage of the gaps that can appear at any time of year. Small-seeded species, for example, are obliged to germinate under moist conditions so that their seedlings do not desiccate. Such species normally germinate during wet seasons. A large-seeded species, however, might be able to germinate successfully in the dry season by drawing on the greater moisture and energy reserves contained in the seed. A seedling that became established before the onset of the wet season would gain a head start and enjoy a competitive edge over species that germinated later. This possibility was examined on Barro Colorado Island by Nancy Garwood while she was a University of Chicago graduate student.

Her results have definitely resolved the issue. Although seeds fall to the forest floor on BCI throughout most of the year, germination is almost entirely suppressed during the dry season. Large numbers of seeds remain dormant in the leaf litter until the first rains have soaked the ground. Both long-dormant and freshly fallen seeds then sprout simultaneously in a veritable orgy of germination. Few if any species exploit the change in seasons. Short-term temporal heterogeneity thus does not make any apparent contribution to tropical tree diversity on BCI.

At longer time scales, it can be asked whether rare events such as fires, droughts, floods, earthquakes, and wind storms play a role in the dynamics of tropical forests. The answer is certainly yes, as a number of well-documented examples have demonstrated. Yet tropical trees do not have annual growth rings that can be used in dating, and so investigators cannot search for telltale evidence in the form of even-aged cohorts of trees. Even when the date of a fire or hurricane can be firmly established from newspaper accounts, evidence that the event led to a burst of recruitment is almost impossible to obtain. In temperate forests, in contrast, the history of any stand can be deduced by coring a sample of trees and counting rings.

This past summer I had the opportunity to visit an 800-year-old Douglas fir forest in the Oregon Cascades. Many of the firs had been standing since the establishment of the forest, while none were older, and only a few were appreciably younger. Douglas firs do not develop in the shade of adults, and regenerate from seed only after major "stand-initiating events," normally fires. Accordingly, after nearly a millennium this stand was still undergoing succession. The lower and middle zones were filling with western hemlock, a slow-growing, shade-tolerant species that in several hundred years would come to dominate the site, as the ancient firs finally died and toppled over. It is impressive to realize that an event 800 years earlier had contributed to the diversity of this forest by allowing the establishment of Douglas fir.

To what extent do rare stand-initiating events contribute to the diversity of tropical forests? In most cases this question cannot be answered, although a gradually accumulating body of evidence suggests that tropical forests may be influenced by rare catastrophic events more widely than had been previously imagined.

Many authors had asserted, for example, that evergreen tropical forests were impervious to fire. Confidence in this belief was shattered by the unusually strong El Niño climatic event of 1983, which brought 2 meters of rain to the normally rainless Peruvian coastal desert and an unprecedented fire to the normally rainy island of Borneo. Fires started on that island by slash-and-burn cultivators began to spread into uncleared forest, especially where slash had been left behind after selective logging. The fires expanded and merged until much of eastern Borneo was smoldering. Over 300,000 square kilometers were damaged or destroyed before the fires were finally extinguished by renewed rainfall. It is too early to say what lingering marks this episode will leave on the vegetation, but surely they will be pronounced, for the mortality of canopy trees in many areas was extensive.

The same 1983 El Niño that brought drought to Borneo also brought drought to Panama. The biological effects of this occurrence might have gone unnoticed had it not been for the alert team of scientists monitoring the 50-hectare plot on BCI. A 1985 recensus of the plot revealed a major spasm of tree mortality. Trees in the understory survived better than those in the canopy. Most severely affected were the largest trees, those exceeding 64 centimeters in diameter at breast height. A surprising 15.5 percent of these trees died. Here was an event of major significance. Not only was 15 percent of the canopy vacated in a single incident but most deaths occurred in species that are most abundant in the wetter forests of the Caribbean drainage of Panama. Species that are more common in the dry forests of the Pacific coast were affected little or not at all. This one event may well have repercussions that will persist for a century or more.

It has recently been discovered that almost anywhere one digs a pit in Amazonia, there will be thin lines of carbon in the soil profile, clear evidence of past fires. Radiochemical dating reveals that the carbon deposits were formed within the period of probable human settlement, suggesting that at one time or another slash-and-burn agriculture has been practiced over much of the basin. Whether any of the fires were of natural origin cannot be ascertained.

Other types of rare events can abruptly alter the composition of a forest. Forests in the Antilles, Solomon Islands, and elsewhere are periodically ravaged by hurricanes. Hilly terrain in Panama can experience massive slumping during earthquakes. Once-in-a-century floods can devastate the vegetation in river valleys. The origin of even-aged ohia forests in Hawaii can be traced to volcanic eruptions, as this species is among the first to colonize lava flows.

A past disturbance may prove to be the key that unlocks the mystery of some strangely undiverse tropical forests, extensive areas dominated by the "cipo" tree (*Cavanillesia*) in the Darien region of eastern Panama, for example, or the curiously uniform *Gilbertiodendron* forests of central Africa. These forests consist of nearly monotypic stands of a dominant species, surrounded by forests of normally high diversity. Glaring anomalies, such forests constitute one of the puzzles of tropical ecology.

The pace of recovery from large-scale disturbance may be unexpectedly slow, as is dramatically illustrated by the forests around Tikal, Guatemala, and other sites in lowland Central America that were abandoned by the Maya civilization 1200 years ago. The diversity of these forests is anomalously low, even in comparison

A stand of ohia trees *(Metrosideros polymorpha)* in Volcano National Park, Hawaii.

to other forests in the same regions, and many of the common species are ones known to have been cultivated or used by the Maya. Even after a millennium, plant diversity in these formerly settled areas seems not to have fully recovered. Let this be a retort to those who glibly assert, in justifying their own destructive designs, that once felled, the "jungle" springs up again almost overnight.

These stories of droughts, fires, hurricanes, and landslides are hardly more than anecdotes,

Some tropical forests are dominated by just one or a few tree species and contain hardly more diversity than some temperate forests. Most low-diversity tropical forests are found where there are unusual soil conditions, such as in this swamp in Costa Rica.

and at this point they do not contribute in a major way to a scientific overview of the dynamics of tropical forests. But as such rare events occur, scientists are eager to study them, and several of the singular occurrences mentioned above are now the subjects of long-term studies. Eventually such studies will lead to a clearer understanding of the role temporal variability plays in maintaining the diversity of tropical forests.

The Intermediate Disturbance Hypothesis

These accounts of rare catastrophes lead us to what is known as the "intermediate disturbance hypothesis," proposed by Joseph Connell of the University of California at Santa Barbara. While studying seedling survival and establishment in the rain forest of Queensland in northeastern Australia, Connell observed that many species seemed to owe their presence to past disturbances. Under the conditions prevailing during Connell's research, these species were not reproducing, and their populations were progressing toward senility. In the absence of another major disturbance, they would eventually disappear from the stand.

Connell's hypothesis has an intuitively appealing ring. In the absence of disturbance, interspecific competition will run its course, and, over time, one or a few species will come to dominate any stand. At the other extreme, frequent severe disturbance will select for a few fast-growing, precociously reproducing species. Slower-growing species would disappear because they would not reach reproductive maturity before the next disturbance eliminated their populations. This is what happens in cultivated fields. What we call "weeds" are plants that possess life histories synchronized to the cropping cycle.

Although Connell's hypothesis can be applied to a wide range of natural systems, its initial intent was to explain how diversity might be maintained in forests. Presented in terms of extremes, as in the above paragraph, it is almost a truism. Some intermediate and variable rate of disturbance will inevitably accommodate the greatest diversity because conditions at different times will favor species having a variety of life history adaptations. A disturbance that occurred with metronomelike regularity would not have the same effect because it would select for species whose life cycles, like those of weeds, were synchronized with the period of the disturbance.

Although the intermediate disturbance hypothesis was inspired by Connell's study of the Queensland rain forest, it has not yet been possible to test it with data from tropical for-

A virgin savannah of longleaf pine and wiregrass in Georgia, United States. The species composition and open character of this plant association are maintained by frequent fires. This type of savannah may have occupied between 15 and 25 million hectares in the southeastern United States before European settlement, but has been reduced to a few scattered remnants by logging and fire suppression.

ests. What is critical to Connell's mechanism is not the absolute frequency of disturbance, but how much the intensity and rate of disturbance varies. This presents a serious challenge to the investigator, who must document a large number of rare events to obtain an accurate picture of the pattern of disturbances. The difficulty of obtaining long-term records has meant that Connell's hypothesis has remained largely an appealing abstraction.

Distance Dependence Revisited

In his study of the Costa Rican dry forest, Hubbell found that as the number of adults per hectare increased, the number of juveniles per adult decreased, at least in a majority of the 30 most abundant species. When Hubbell repeated this exercise using data from the 50-hectare evergreen plot on BCI, he found much weaker indications that a high density of adults led to depressed numbers of juveniles per adult. Only one species, *Trichilia tuberculata*, the most common canopy tree, showed this negative relationship to a strong and significant degree. The evidence seemed to indicate that the presence of adults had only a weak and infrequent impact on the recruitment of juveniles, apparently too weak to limit the size of adult populations. In assessing the results, Hubbell reaffirmed his earlier conclusion that non-equilibrium processes regulate the species composition of tropical forests. The evidence that species densities fluctuated around an equilibrium was not sufficiently strong and pervasive to be convincing.

Following the publication of these results, and five years after completing the initial mapping, Hubbell and Foster carried out a recensus of the entire BCI 50-hectare plot. Among their first findings was evidence that the growth and survival of saplings was markedly reduced when the nearest neighbor was an adult of the same species. In 11 common species, three-year survival was reduced by 5.8 percent, and in "rare" species as a group, by 9.6 percent. The saplings grew at rates that were 17 and 22 percent slower, respectively. Such drops in performance were found not only in small saplings, but in larger juvenile trees as well. The revised analytical approach—examining the interactions between neighboring individuals instead of seeking correlations within whole subplots—reveals the spatial scale at which negative interactions operate, and explains why such weak evidence had been obtained earlier when the sampling units were 1-hectare subplots.

These new findings of Hubbell and Foster now provide unambiguous evidence across a broad selection of species that distance-dependent forces can cause declines in the next generation. It has yet to be shown, however, that the observed effects are strong enough to prevent common species from excluding rare ones over time. This remains an important missing piece in the puzzle.

The work of Hubbell and Foster has given us a greatly expanded understanding of the processes that regulate tree species diversity in tropical forests. Before results began to emerge from the plots in Costa Rica and Panama, the whole subject was a mystery. Now, we have at least a dim light guiding the way as we grope toward a fuller understanding.

Diversity among Trees: Where Are We Now?

This chapter has pursued the question of why there are so many species of trees in the tropical forest. In trying to find answers, we have examined several hypotheses proposed by workers in the field. To arrive at a final synthesis, let us briefly review each of these hypotheses and what the evidence has shown us about it.

Tree fall gaps of different size and compass orientation may be colonized by different tree species. Although it is not clear that the gaps that appear in tropical forests offer more diverse opportunities for recruitment than gaps in temperate forests, the very large size of some tropical gaps suggests that they might do so. Different species may be found in the root, trunk, and crown zones of fallen trees, suggesting that tropical microsuccessions may be more complex, but comparable data from the temperate zone are needed to confirm whether this is so. It also seems likely that tropical tree communities include species adept at regenerating in zones of arrested succession where vines or bamboo have suppressed other species. Trees that were adapted for establishment in vine-choked openings or bamboo patches might constitute an exclusively tropical "guild." It is doubtful, however, that such a guild would contribute more than a handful of species to the overall diversity of a forest. In general, tropical forests seem to offer a somewhat more varied mosaic of recruitment conditions than do temperate forests, but the likely difference falls far short of that needed to account for the "extra" diversity of tropical forests.

As for temporal heterogeneity, the work of Nancy Garwood ruled out the possibility that species could reduce competition for establishment by staggering seed germination throughout the wet and dry seasons. There remains the untested possibility, however, that seed germination may be staggered throughout the year in nonseasonal everwet forests. It seems likely that, on longer time scales, the disturbance of fire or drought might create the conditions essential for some species to become established, but again, this is a matter that requires further investigation. Fire is probably a major controlling factor in the ecology of temperate forests, but limited evidence suggests that evergreen tropical forests rarely burn without the encouragement of human beings. Landslides are likely to be more important in the tropics where soils become waterlogged in the rainy season. Severe wind storms may occasionally occur in temperate regions and in the tropical hurricane belt. Other parts of the humid tropics are relatively calm. Ice storms, a peculiarly midlatitude phenomenon, can do spectacular damage to temperate forests. This one is hard to judge; there is no clear winner.

Hubbell's non-equilibrium hypothesis proposes that tropical forests are composed of competitively equal species in a state of non-equilibrium. If the populations of species are fluctuating at random, then no two plots should have the same species present in the same abundances. This prediction is contradicted by Ashton's results on tree communities in East Malaysia, however. On the scale of nearest neighbors, Hubbell's hypothesis is contradicted by his own results showing that juvenile trees do not grow or survive well when located next to mature adults. The failure of its predictions renders this hypothesis less attractive than alternative hypotheses. Moreover, there is no evident reason why the absence of density (distance) dependence would necessarily lead to more diverse tree communities in the tropics. Hubbell's proposal is rapidly losing its appeal, as he and Foster obtain stronger evidence of the widespread operation of distance-dependent effects in tropical tree populations.

Connell's intermediate disturbance hypothesis provides a powerful insight into the perpetuation of diversity in both temperate and tropical forests. The size range of tree fall gaps is larger in the tropics, but catastrophic disturbance may be less frequent than in the temperate regions. Whether the opposing effects of these trends tend to cancel out one another is unknown.

In addition to the spatial and temporal variation produced by the vagaries of tree falls and the weather, we have examined a number of biological mechanisms that promote diversity. Hubbell and Foster found that saplings on BCI display affinities for diverse light conditions, and they distinguished "guilds" of species whose juveniles were found in full sun, partial sun, or shade, or appeared "indifferent" to the light environment. It is possible, although still unproven, that tropical forests could support a greater number of such regeneration guilds than temperate forests by offering a wider range of microsites in which establishment could occur.

Last, but not least, is the possibility that strong distance-dependent forces, mediated through seed predators, fungal pathogens, or herbivores, could limit the abundance of tropical tree populations, thereby reducing competi-

An agouti (*Dasyprocta aguti*) hunts for seeds in French Guyana. These large rodents are especially important seed predators and dispersers throughout the Neotropics because of their habit of scatterhoarding (hiding and burying) surplus seeds. Once buried, the seeds are protected from other seed predators, and their chances of survival are improved because the agouti often forgets to retrieve them.

tion and permitting the coexistence of many species. These effects are likely to be stronger and more common in the tropics because tropical forests harbor more seed predators, fungal pathogens, and host-specific herbivores than temperate forests. The operation of strong negative distance effects in itself, however, does not necessarily lead to a predictable community structure such as Ashton found in East Malaysia. Important mysteries remain to be solved before a full understanding of tropical tree diversity is within our grasp.

We have now come full circle in the arguments. In the last chapter, we concluded that tropical forests support more species of birds, in part because the high diversity of plant species ensures a richer smorgasbord of food resources than is available in temperate forests. Now, in this chapter, we have concluded that the greater diversity of animals in the tropics may enhance the diversity of plants through the effects of seed predators, pathogens, and herbivores. The arguments, although seemingly circular, are legitimate because the interacting plant and animal species are ultimately created by the processes of evolution (to be considered in Chapter 6). Once new species have arisen, they continue to adapt. Many evolve complex interdependencies with other organisms that share the same environment. It is thus impossible to consider animal diversity independently of plant diversity, or vice versa. These two main components of biodiversity are inextricably bound together in both evolution and ecology.

It should now be clear that the subject of diversity is enormously complex, compelling the investigator to consider processes that operate on vastly different scales of time and space. Even now we are only halfway through our quest for answers to the question, Why are there more species in the tropics? In this chapter, the inquiry has been directed toward what may be termed the "horizontal component" of plant diversity. That is, we have been concerned with how diversity increases with the area of sample plots and with how nearest neighbors may affect one another. Our arguments have almost ignored the fact that nearly every plant in a tropical forest grows above or beneath other plants. There is consequently a vertical component of diversity in addition to the horizontal one. This vertical component will be the topic of the next chapter. Beyond that, we shall turn to the evolutionary dimension, certainly the most fundamental of all from the standpoint of ultimate causality.

Sunlight and Stratification

We earthbound humans concentrate on the horizontal. Walking through a forest we look ahead or at the ground to ensure solid footing, but seldom do we look up. In not doing so we fail to appreciate the quintessentially three-dimensional environment. Many tropical forests attain heights of 40 or 50 meters and, in Southeast Asia, even 60 or 70 meters. Confined as we are to the bottom 2 meters of this arboreal world, we have direct access to only about 5 percent of the volume of the habitat, and the least productive 5 percent at that. A terrestrial mammal in this environment is in the position of a crab on the floor of the sea: everything important comes from above—not only the sun and the rain, but more than three-

Light beams into the forest interior
through gaps in the canopy, supporting
the photosynthesis of subcanopy and
understory trees.

Day

Day and night

Trees

Giant squirrel

Red leaf
monkey

Bornean gibbon

Orangutan

Trees and ground

Yellow-throated marten

Sun bear

Plain
pigmy
squirrel

Long-tailed
macaque

Slender tree shrew

Binturong

Ground

Tufted ground squirrel

Muntjak

Short-tailed
mongoose

Malay weasel

Sumatran rhinoceros

Asian elephant

White-toothed shrew

Large tree shrew

Bearded pig

Lesser mousedeer

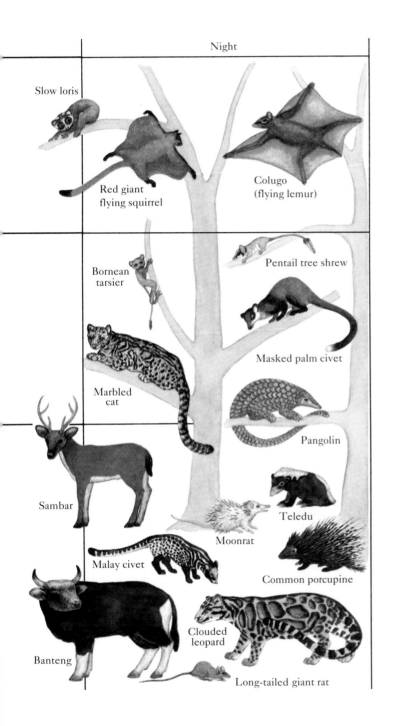

Night

Slow loris

Red giant
flying squirrel

Colugo
(flying lemur)

Bornean
tarsier

Pentail tree shrew

Masked palm civet

Marbled
cat

Pangolin

Sambar

Teledu

Moonrat

Malay civet

Common porcupine

Clouded
leopard

Banteng

Long-tailed giant rat

quarters of the food supply that sustains the quite ample community of terrestrial birds and mammals. Many of these animals subsist on plant parts that fell before they could be eaten by canopy-dwelling species; most of the rest are carnivores.

Much that is important to the functioning of the forest happens in the canopy, the upper tiers of foliage arrayed 20 to 50 or more meters above the ground. The photosynthetic activity of the forest is concentrated in this zone, where the production of fruits, leaves, seeds, and nectar is far higher than in the lower tiers of vegetation. In response to the high productivity of the canopy, a wide range of partially to wholly arboreal mammals has evolved, as mentioned in Chapter 4. In the Neotropics alone there are leaf-eating sloths and howler monkeys; twig- and bud-eating porcupines; prehensile-tailed anteaters; nectar-feeding opossums and kinkajous; a large, bamboo-eating rat; monkeys that feed on fruits, leaves, seeds, or insects; as well as the sap-eating pigmy marmoset, the primate counterpart of a sapsucker. Arboreal quadrupedal mammals, joined by birds and bats, consume nearly every imaginable product of the canopy, save the wood itself. Yet, except for the primates, which are large, diurnal, and easily habituated to observation, most of these animals have been little studied. The canopy and

Some mammals of a forest in Sabah, Borneo. About half the species are arboreal, and half terrestrial. Diurnally active species are portrayed on the left; nocturnal species on the right.

(Left) A common iguana *(Iguana iguana)* basks in the branches of a *Cecropia* tree in the New World tropics. Cold-blooded iguanas accelerate the process of digesting foliage by warming their bodies in the sun. *(Right)* A prehensile-tailed porcupine *(Coendou mexicanum)* browses in a Costa Rican treetop. Although widespread in tropical forests around the world, porcupines are shy, nocturnal, and difficult to observe.

its animals are largely shielded from human inquisitiveness by intermediate layers of foliage and the difficulty of access. Having trod on the moon, modern man has still not devised a convenient means for investigating the roof of the forest.

Organization in the Vertical Dimension

The horizontal expanse of the forest floor is a patchwork of recent and past tree falls. Not surprisingly, it gives an impression of relentless disorder. In its vertical dimension, however, the forest is organized in a more orderly fashion. Broad-crowned emergents tower above the general canopy, often reaching heights over 50 meters. Such giants seldom form a continuous canopy; more often they are widely scattered, separated by broad gaps through which sunlight reaches the tiers of foliage below. The densest foliage commonly lies under the emergents, frequently at heights between 20 and 30 meters. Many species of trees coexist at this level, and most of them possess crowns considerably narrower than those of the emergents. Below this second tier, other, small plants take their places in the vertical sequence—treelets, shrubs, and finally herbs. Small plants tend to occur near the ground, and their crowns become broader as they grow taller.

There is a simple explanation for this trend. As a plant grows taller, it must make an ever increasing investment in roots, branches, and a trunk, which all contain living tissues that respire (recall Chapter 2). To support the increasing cost of respiration borne by an expanding superstructure, a plant must continuously increase the amount of foliage it displays to capture sunlight. Most plants do this by expanding the size of their crowns. However, growing plants are subject to certain physical constraints that limit the range of shapes they can assume.

Thomas McMahon of Harvard University has shown that the dimensions of trees conform to a predictable allometric rule. Given the known tensile and shearing strengths of wood, it can be proven that trees of any given height must possess trunks of a minimum diameter in order to avoid buckling under their own weight. Many species have been tested, and all conform to the prediction with room to spare. The diameters of even the most slender individuals are always greater than the minimum needed to support the trunk. The extra girth presumably provides trees with a margin of safety so that they won't snap in the first gust of wind.

The order apparent in the vertical plane of forests is a consequence of this design constraint and a counterbalancing constraint derived from the tendency of plants to allocate their resources efficiently. Trees are physiologically obliged by the nature of the growth process to continue increasing in height and girth throughout life. However, growth rates vary markedly over the life span. Young trees are at severe risk of being shaded by others, so all available resources are allocated to increasing height within the mechanical constraint described above. A need to minimize the risk of being overtopped by a neighbor explains why, in an adaptive sense, most trees will not bloom and bear fruit until they are at least several years old.

Once a tree has attained a height that ensures a sufficient supply of light, it can reorder its priorities to favor the production of fruits and seeds over continued growth. At this point, height and girth increase at a diminishing and sometimes almost imperceptible rate. Trees do not waste resources becoming larger than they need to be. The rare individual of monumental girth is thus likely to be extremely old. Because forests composed of adult trees put their resources into reproduction instead of wood, "old growth" forests are anathema to foresters. The derogatory term "overmature," employed by the timber industry in referring to such forests, is a cynical misnomer, designed not for its representation of fact but for its public relations impact.

In general, the taller the tree, the larger the crown needed to pay the cost of maintaining the trunk, branches, and root system. Large crowns do not form on trees of low stature in the understory because growth rates are inherently very low in the shade. An understory tree that deferred reproduction until it had built a large crown might be smashed by a falling trunk before it produced seed. Moreover, the understory is crowded with countless small trees and saplings, so the necessary space for expansion is not available. In the presence of these constraints, tree crowns expand with height in a notably regular fashion.

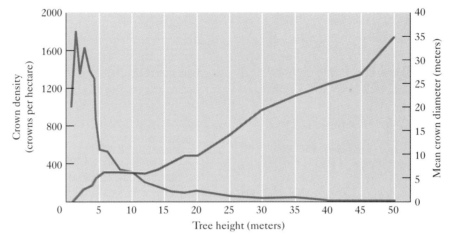

In moving upward through a mature Peruvian floodplain forest it is apparent that the density of crowns decreases while their mean diameter increases. The tallest emergents spread their branches over a tenth of a hectare, shrouding a hundred lesser trees in their shade.

The canopy and understory consequently present very different environments. At a height of 40 meters in the canopy of the forest I study, about 60 percent of the space in the plane is occupied by tree crowns. The average crown at this height is 25 meters in diameter, so it shades an area of nearly 500 square meters. Only 12 such crowns occur in a hectare. Two meters above the ground in the understory, the situation is radically different. The average crown projects over an area of only 2 square meters, and is merely one of more than a thousand that occupy a hectare. In between the two levels, trees are intermediate in number and crown size.

These structural differences are not the only ones of biological consequence. The huge crowns and high species diversity of the canopy mean that a given tree may be tens or hundreds of meters away from its nearest neighbor of the same species. From the perspective of an animal that exploits the canopy, whether pollinator, frugivore, or phytophageous (leaf-eating) insect, the canopy is an extremely patchy environment. Perhaps for this reason, songbirds that inhabit the canopy frequently occupy territories four to five times larger than their understory counterparts. To exploit the patchy environment successfully, they must move over large distances.

The much smaller crowns of adult plants in the understory, and the often reduced species diversity of this stratum, carry a different set of implications. Crowns in the canopy are typically discrete, being separated by gaps of a few to several meters. These gaps probably serve to reduce mutual shading, but may also prevent damage from the whipsawing action of adjacent crowns during wind storms. In the dif-

Propelled by long, powerful hind legs, this sifaka (*Propithecus verreauxi*) and related Madagascar primates are able to move quickly from trunk to trunk by cling-and-leap locomotion. In other parts of the tropics, primates may navigate through the forest by jumping, brachiating, or walking on vines and branches. A few of the largest species, such as baboons, gorillas, and orangutans, prefer the ground for long-distance travel.

fuse light and still air of the understory, the branches of adjacent treelets commonly interdigitate so that the foliage is almost continuous. Walking animals can therefore move easily from one plant to the next without having to move up or down or to make spectacular leaps. Plants of a given kind are much closer together than in the canopy, but they are individually small, so the resources available in each are

limited. Because lightly constructed plants cannot support much weight, the birds and mammals that inhabit this zone of the forest are invariably small. In the middle and upper tiers of the forest, the branches are stout and can resist the weight of larger animals. But even in the upper canopy where the branches are stoutest, it is rare to find animals weighing more than 12 kilograms. The 80-kilogram orangutan is thus an extreme oddity that should not exist but somehow does.

In addition to these structural distinctions between canopy and understory, there are pronounced microclimatic differences that are of consequence to plants and animals. Exposed to the full sun and open air, the top of the canopy is hot, relatively dry, and often windy by day. The loss of heat radiating upward on clear still nights can sharply lower the temperature at treetop level. In contrast, the environment near the ground is perpetually humid, poorly illuminated, and practically windless. The daily swings in ambient temperature at this level are significantly less than in the canopy.

Because they differ so in both microclimate and physical structure, the canopy and understory are regarded as distinct domains by many of the forest's animal inhabitants. Numerous mammals, birds, reptiles, and arthropods spend their entire lives within one vertically circumscribed zone, thereby leaving other zones available for exploitation by different species. A number of questions therefore come to mind. Are the vertical zones occupied by animals broader in the understory or in the canopy? Do more species coexist at one level than another? Do the dietary habits of consumers depend on the vertical zones they occupy in the forest?

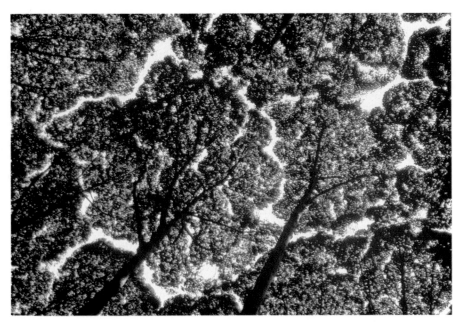

Kepong trees *(Dryobalanops aromatica)* in this Malaysian dipterocarp forest exhibit the phenomenon of crown shyness. The branches of adjacent trees avoid interdigitating, leaving gaps through which sunlight can penetrate into the forest interior.

The only study to examine these questions explicitly was conducted with birds at a locality in the foothills of the Peruvian Andes. At 700 meters elevation, the site supported fewer species than the Amazonian lowlands, about 156 in all. Over several visits, I recorded data on the foraging heights of 134 species. The remaining 22 species were raptors, vultures, night birds, and aerial feeders to which the measurements did not apply. The foraging zone of each species was defined as the mean plus or minus one standard deviation, a range that includes about two-thirds of the observations. This tactic allowed me to tally the number of species whose foraging zones overlapped at any height in the forest.

The results clearly showed that the density of species is higher—nearly twice as high in fact—in the lower and middle portions of the canopy than below a height of 5 meters in the understory. To understand this result, we must first consider some of the contributing factors.

The depth of any species' foraging zone is ultimately constrained by the height of the forest. Species living near the ground or in the topmost canopy have less latitude to vary their foraging heights than those living midway between the two. Accordingly, species occupying the middle tiers of the forest exhibited broad foraging zones compared with those living near the ground. Some of the ground-dwelling birds occupied extraordinarily narrow zones only a few meters deep.

How much a species concentrates its foraging activity in the vertical dimension is measured by the depth of its foraging zone—the narrower the zone, the more concentrated its activity. Thus, at any level it is possible to estimate the intensity of foraging activity by all species together: we divide the number of species whose foraging zones overlap at that level by the mean depth of their foraging zones. Performing this calculation, we find that the forag-

ing activity of all species together is nearly the same at all heights. Thus, birds do not seem to be using one level more intensively than others, despite the large variation in the number of species present at different levels.

There must be other causes for the differing numbers of species at each level. It is not clear what these are, although the relative paucity of species in the understory has a number of possible explanations. No species whose foraging is centered below 10 meters weighs more than 100 grams. Plants at this level are not

The numbers of bird species whose mean foraging positions fall within the indicated height ranges. The scene represents an Amazonian forest in Peru.

able to support much weight, but perhaps more importantly, the greatly reduced photosynthetic productivity of the shady understory may be insufficient to meet the metabolic demands of a large consumer. Several major guilds are virtually absent from this zone, including mast feeders, trunk foragers, and the superguild of omnivores. The members of these guilds are instead concentrated in the canopy. Understory birds tend to be dietary specialists, feeding on fruit, nectar, or insects, but not on combinations of these resources.

Nearly all the omnivorous birds of the forest occupy the canopy. Why? An educated guess is that omnivory is an evolutionary response to the extreme patchiness of the canopy environment. Canopy crowns are large and far apart, as was noted above, and moving between them requires a significant expenditure of energy. Now, imagine a specialist consumer that ate only fruit, nectar, or insects. All these resources are produced in the canopy, but by particular tree species at particular times. It is likely that even insects are patchily distributed in the canopy, since they would be most abundant in crowns that are producing new foliage. The canopy is thus a boom-or-bust environment. A bird that ate only fruit, for example, might have to fly long distances between the few individual trees that were producing fruit at a given time. In contrast, a less selective dietary generalist might find many more trees that offered sustenance, and these trees would be closer together. In this situation, the generalist wins.

Why do generalists not enjoy similar advantages in the understory? Individual plants are much smaller in the understory, and most

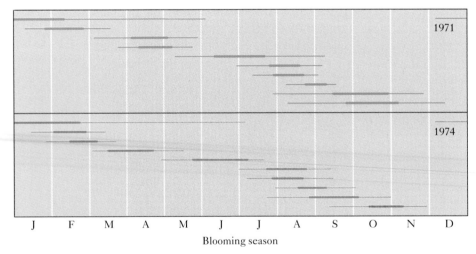

The blooming seasons of *Heliconia* species in a Costa Rican rain forest are staggered, ensuring that nectar is always available. It has been proposed that the flowering periods of these banana relatives have diverged to reduce competition for the services of their hummingbird pollinators. Note the similar order of flowering in two years.

A hermit hummingbird (*Phaethornis superciliosus*) receives a load of pollen on its forehead as it drains the nectar from a *Heliconia* flower. The yellow flowers are made conspicuous by the bright red bracts.

species are represented by many more individuals per hectare than is the case of canopy trees. Many species, perhaps because light is scarce, produce fruit and flowers a few at a time on extended schedules. Fruit or nectar is thus more reliably present in a given location. Given a slow, stable production of resources, and a high predictability of finding them, populations of consumers will increase until their collective demand closely matches the level of production. Under these circumstances, only the most efficient harvesters will realize a sufficient gain to pay the additional costs of reproduction. Here it pays to specialize, for a jack-of-all-trades can be master of none.

Having seen how the vertical organization of the bird community in one Amazonian forest can be related to various features of the plant community, we shall now turn to the plants themselves to explore how they might become stratified into distinct tiers. One can speculate that stratification of the plant community contributes in turn to the vertical organization of the animal community, but the specific mechanisms have not been clearly articulated.

Vertical Stratification: Reality or Fantasy?

There is no question that tropical forests possess vertical organization, but just how and to what degree they are organized is a matter of considerable debate. That the trees of the canopy are larger than those of the understory no

The profile of a lowland mixed dipterocarp forest in Brunei, Borneo, hints at the presence of distinct strata. Dipterocarps tower above the many lesser trees that occupy the forest understory. "Flying" frogs, lizards, lemurs, and squirrels have evolved sail-like structures that allow them to glide from one lofty crown to another in Southeast Asian forests.

one will deny, but beyond this most preliminary generalization there is little agreement. Some authors see definite order; others see only chaos. Is the truth simply what is in the eye of the beholder?

The most renowned advocate of order is Paul W. Richards. Early in his career, Richards spent several years studying forests in (then) British Guiana, Nigeria, and Malaysia. His efforts culminated in the publication in 1952 of a landmark book on tropical forests, which to this day remains a valuable sourcework. Richards expounded the thesis that tropical forests

throughout the world are vertically stratified. He proposed that the strata are comprised of distinct suites of plant species, each intrinsically adapted for the conditions at a particular level. In all, he recognized five strata and termed them A, B, C, D, and E, starting with the towering emergents and working down to the lowly herbs on the forest floor. The strata were held to be independent of one another in the sense that all of them would be routinely encountered over randomly selected points on the ground, except, of course, where there were recent tree falls. Imagine that a giant could poke needles down through the forest from above. Richards's claim was that the needles would pass through an average of five plant crowns before reaching the ground.

Reactions to Richards's thesis were mixed. Some authors, especially textbook writers, reacted positively. Here was a wonderfully heuristic scheme that provided a focal point for lectures about the rain forest. Since most biology textbook writers have never seen a rain forest, it wasn't a matter on which they held strong opinions. Responses from tropical biologists were mostly guarded or negative. Some looked but failed to see. It wasn't clear how to prove the idea, or what precisely should be measured. Moreover, the formidable challenge of identifying rain forest trees inhibited ecologists from making a concerted study of the issue.

After enduring years of criticism from his colleagues, and in the absence of positive support from any source, Richards reluctantly came to admit that the strata "may have no objective reality." Although the issue remains unresolved to this date, strong evidence has recently appeared that some temperate forests are stratified in just the manner Richards imagined. What was lacking in Richards's proposal was an explicit mechanism that would account for stratification and that was amenable to rigorous quantitative test. Such a mechanism is now available.

The Sunfleck Model

Temperate forests are much simpler than tropical forests, and it is in the simplest cases that science often finds the answers to more difficult questions. South of about 42 degrees of latitude in the eastern United States (central Wisconsin, northern New England), a number of shade-adapted woody species appear in the mid-story of deciduous forests. Some of these, such as flowering dogwood and redbud, are conspicuous and well known to everyone, while others, such as hophornbeam and spicebush, are more obscure.

A number of years ago I noticed that these species attain a certain height in a given forest and then cease their upward growth. Unlike canopy species that pursue an upward trajectory until they reach the open sky or die, dogwoods and the other members of this mid-story "guild" are adapted to live out their lives in the shade of the forest interior. Specimens of large girth are no taller on average than slender individuals a fraction their age. If one climbs a tree and looks out through a still, leafless spring forest when the dogwoods are in bloom, it is striking to note how the blossoms all lie in a single narrow plane some 7 to 10 meters above the ground, well below the lowest limbs of the canopy.

These observations suggested that there might be an optimum height for a mid-story tree and that dogwoods and other similarly adapted species could be responding to a signal in the environment that indicated the level of this optimum. The hints that there might be such an optimum were strongly reinforced when I discovered that, in forests containing two such mid-story species, they both converged on the same limiting height. The two species were apparently responding to the same signal in the environment, but there was no clue as to what the signal might be.

An extremely mundane observation provided the key to the puzzle. Flowering dog-

Flowering dogwoods celebrate the arrival of spring in an oak-hickory forest in central New Jersey. The dogwood flowers are displayed in a narrow plane well below the forest canopy. Colonies of may apple (*Podophyllum peltatum*) contribute a herbaceous layer to the three-tiered structure.

woods are commonly grown as ornamental plants in yards and gardens. In such settings, dogwoods are lower and bushier, and flower and fruit more profusely, than their forest counterparts. The abundant sunlight available in gardens seemed not only to increase reproductive output but to markedly affect the shape and height of the plants as well. It seemed possible that dogwoods growing in their natural environment were restrained by lack of light.

I became curious as to how this species and other members of the mid-story guild received sunlight from their vantage point well beneath the crowns of much taller trees. Previous studies had shown that about 30 percent of the light reaching the ground in eastern deciduous forests came in the form of "sunflecks," shafts of direct beam light that entered through holes in the vegetation above. Might it be possible to calculate how sunflecks contribute to the light received at different levels under the canopy?

It turned out that the calculation is really quite simple if one takes the liberty of assuming that the canopy is a regular array of close-packed crowns. From the perspective of an observer on the ground, the canopy is like a plate dotted with holes through which direct sunlight can pass. But the plate is not a flat one; it has a thickness equal to the depth of the crowns in the canopy. If one looks at a forest profile, along the edge of a cultivated field for example, it is easily noted that the crowns are not cylindrical, but peaked, with the highest point in the middle. The gaps between crowns are thus funnel-like in cross section. Because gaps are wider at the top than at the bottom, sunlight penetrates into the forest over a wide range of angles.

When the sun is low in the sky, direct beam illumination does not pass into the forest interior at all. Once the sun has risen just high enough, direct light passes through the funnel-shaped gaps into the forest interior, and a sunfleck forms to the west of the gap. As the sun continues to progress across the sky, the sunfleck moves eastward, slowing at midday and accelerating in the afternoon, as it circum-

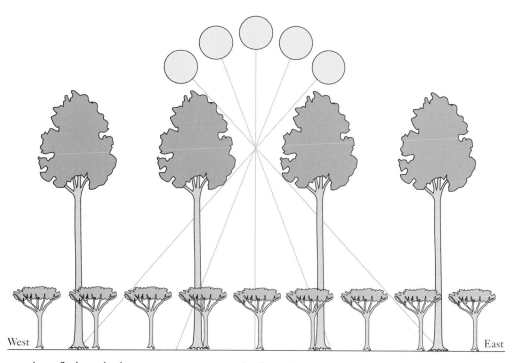

A sunfleck tracks from west to east across the floor of a temperate deciduous forest as the sun passes overhead. The shapes of the crowns, and the widths of the gaps between them, determine the angle at which direct sunlight can enter the forest interior.

scribes a track across the forest floor. Each day prior to the summer solstice the location of the track shifts slightly southward; after the solstice the tracks again progress northward. Over the course of a growing season the forest floor is thus more or less evenly illuminated, although on any given day there will be shaded strips between the tracks of adjacent sunflecks. The equable spreading of these tracks over time presumably explains why temperate forests often support a nearly uniform herb layer despite the extremely patchy distribution of sunflecks at any given moment.

We now have a rough description of the behavior of sunflecks that provides a simplified overview of the light field at the level of the forest floor. In order to understand the distinctive behavior of dogwoods and other mid-story species, we must now inquire into how the light field under a canopy varies with height.

The light admitted into the forest interior by a single gap passes through a triangular area on the way to the ground (assuming for the sake of simplicity that there are no intervening layers of foliage). The apex of the triangle lies in the gap, and its base on the ground. A point

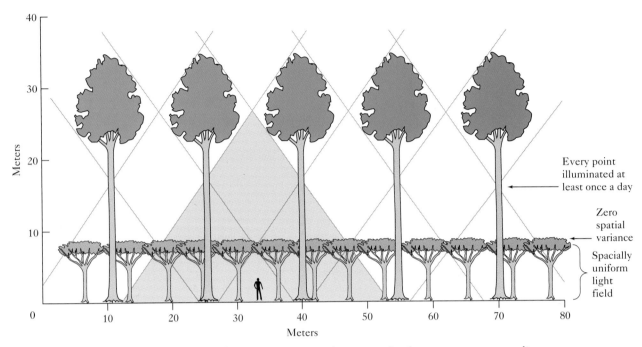

The light passing through a canopy gap over the course of a day generates an expanding cone of illumination in the forest interior. Where the cones of alternate gaps intersect, a horizontally uniform light field is produced in the understory. Dogwoods and other mid-story trees grow up to but not beyond this level.

in the narrow portion of the triangle may receive several hours of light, but farther down, as the light is spread over an ever broadening strip, its duration at any spot is proportionately diminished. As the triangles of light from adjacent gaps spread out below the canopy, a point is reached at which the western end of one begins to overlap the eastern end of its neighbor. Points in the area of overlap receive relatively brief periods of illumination twice a day, once in the morning and once in the afternoon. Descending even farther below the canopy, the expanding triangles overlap more and more so that any point receives light from an increasing number of gaps. By doing some rather simple arithmetic, it is possible to calculate the total

light received from sunflecks at any point along a horizontal line at any depth below the canopy. The interesting feature of the results is that the variation from point to point in the amount of light received along these lines is high beneath the canopy and drops nearly to zero where the light triangles of alternate gaps begin to overlap. From this plane downward, the light field within a forest is spatially uniform, though it continues to be temporally variable.

Under these circumstances, what is the optimum height for a plant that is to remain permanently in the understory? In answering this question, I reasoned that a tree must grow at least as high as the upper limit of the spatially uniform light field to avoid being overtopped by competitors. If it grows significantly above this point, it will encounter two disadvantages. First, the light field across its crown will become increasingly heterogeneous, and eventually there will be branches that do not receive enough light to pay their cost. Second, the plant will have to pay higher construction and maintenance costs for the additional trunk and root system needed to support and supply a higher crown. These costs would have to be paid out of resources that might otherwise go to reproduction. The best strategy then is not to grow taller than necessary. To accomplish this, the plant would do best to cease upward growth at the point where it began to detect heterogeneities in the amount of light being received by different portions of its crown. Accordingly, I predicted that mid-story trees should grow as high as the upper limit of the spatially uniform light field, and no higher.

It now remained to test the prediction in a real forest. After much sleuthing, I located stands in Maryland, Virginia, North Carolina, and South Carolina that were mature enough not to be undergoing succession. The mean heights of dogwoods and hophornbeams in the mid-stories of these forests coincided closely with the predicted upper limit of the spatially uniform light field. Affirmation of the predictions gave strong support to the idea that vertical stratification of forests is a genuine natural phenomenon that represents an adaptive response to light conditions under the canopy. But what is the relevance, if any, of these observations on temperate forests to the much more complex issue of stratification in tropical forests?

Richards Revisited: Do Tropical Forests Conform to the Temperate Model?

One might imagine it a straightforward matter to apply the sunfleck model to tropical forests. All that would be required is to carry out the same measurements and procedures used to test the model in temperate forests. Tropical forests, however, differ from temperate forests in some crucial ways that greatly complicate the situation. The sunfleck model assumes an even upper canopy with narrow, more or less regularly spaced gaps. In contrast, the canopy of most tropical forests is notoriously uneven. A-story crowns tend to be widely and irregularly scattered, and the large spaces in between are far wider than the gaps of temperate forests. With the help of a range finder, a strong neck, and some patience one can measure the dimensions of canopy tree crowns from the ground. What cannot be measured so readily is the angle at which light begins to penetrate into the forest interior. Because of the uneven heights of trees in the canopy, and the wide and irregular spaces between their crowns, gaps are apparent to an observer over a broader range of angles than in the temperate forest, and there is no apparent cutoff angle below which no direct light enters. Furthermore, several layers of crowns intervene between an observer on the ground and the upper canopy, and the obstruction from these layers severely hampers measurement of the angular distribution of light available to crowns forming the second highest tier of trees (Richards's B-story). These technical impediments have not been overcome at this writing.

There is, however, a more intuitive approach that provides some hints as to how the stratification mechanism might operate in tropical forests. If one looks at the crowns of trees composing the upper stratum of many tropical forests, they have a distinctly "tropical" look. You would never see such a tree in New Jersey, but you might not be able to say exactly why. The reason is that the trees of the upper canopy have a different shape. They tend to be half again as tall as a temperate tree, and to have broad, shallow crowns. They are more mushroomlike, in other words. More explicitly, the crown of a 35-meter-tall oak in Virginia is about 12 meters in diameter, and 11 meters deep, and it occupies the upper third of the height of the tree. In contrast, the crown of a 55-meter *Dipteryx* in Peru has a spread of 30 meters, twice as broad as the oak crown. However, the depth of the *Dipteryx* crown, at 15 meters, is only slightly greater than that of the oak. If one could transplant the oak and the *Dipteryx* to Fairbanks, Alaska, they would both look ridiculous towering above the puny, slender spruces of that latitude.

There can be little doubt that the crowns of trees are shaped by the angle of solar radiation received during the growing season. In the tropics, the sun passes high overhead every day of the year. Trees that display their foliage over a flat or shallowly domed perimeter are best designed to receive light from directly above. At the latitude of Pennsylvania (40 degrees), the sun reaches a maximum height in the sky of 73.5 degrees on the summer solstice, and, accordingly, midlatitude tree crowns are deeper,

facing outward to receive light from lower angles. Farther north, the sun tends to circle the horizon, so at Fairbanks, near the arctic circle, it never rises above 47 degrees in the sky. To absorb light arriving from low angles, a tree needs to build a deep, steeply sloping crown; hence the picturesque Christmas-tree look of the northern conifer.

The narrow conical crowns and low stature of the boreal forest preclude the formation of a second stratum of woody plants. The attenuated crowns intercept the low-angle sunlight, and very little penetrates to form sunflecks on the forest floor. Deep crowns, narrow gaps, and a short growing season all limit the light available under the canopy. Shrubs may grow where there are gaps in the canopy, but beneath the spruces and firs the forest is dark and nearly devoid of other plants.

Progressing southward to the temperate midlatitudes, the trees become taller, and their crowns assume more compact forms. The growing season is longer, and the sun rises higher in the sky, providing a more generous annual energy budget. Shafts of direct sunlight pass through gaps in the canopy, contributing sufficient light to support a mid-stratum of small trees, as was noted above. Moving farther toward the equator, the space beneath the canopy becomes progressively more filled out with additional plant layers.

Whether mature tropical forests contain five plant strata, as Richards initially claimed, is still an open question. Certainly a midstory of dogwood and hophornbeam has "objective reality" in the simplified structure of a forest in Virginia. This reality loses its clarity in the tropical forest, where there may be as many as three intermediate strata instead of one. An al-

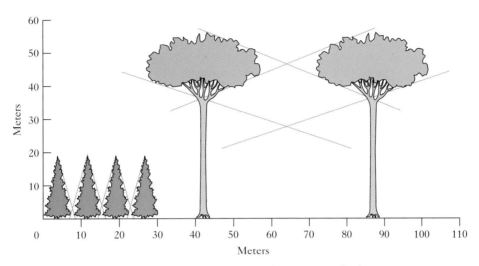

The shallow crowns of tropical forest emergents admit sun into the forest interior over a much wider range of angles than do boreal conifers, allowing additional strata of trees to grow under the canopy.

ready confusing situation becomes even more complicated by the fact that every individual tree that attains a position in one of the upper strata must perforce grow up through all the lower strata. The impression will be one of un-resolvable chaos if an investigator is unable to recognize the species at sight, and if there is no information available on the heights at which particular species reach maturity, as is true for nearly all tropical forests. Richards's critics have not proven him wrong; they have simply pro-tested that his evidence lacked the necessary rigor to fully substantiate his claim. In this complaint they have a valid point.

My own work on this question has not yet reached the stage at which it can conclu-sively vindicate or refute Richards's thesis. Nev-ertheless, evidence obtained in collaboration with Robin Foster, Kenneth Petren, and Jeffrey Mathews provides strong support for some components of Richards's argument, namely that most points on the ground are overtopped by about five superimposed crowns, and that the shapes of crowns vary systematically with their vertical position. What has not yet been demonstrated unequivocally is that the superim-posed crowns are organized into discrete layers, although some of the evidence suggests that this may indeed be the case.

One bit of suggestive evidence comes from plots we have analyzed in Amazonian Peru. All trees greater than 10 centimeters in diameter at breast height were mapped and identified to species. Using a range finder, we measured the heights of the tops and bottoms of all the crowns and estimated their diameters by pacing off their projections on the ground. This infor-mation can be processed in various ways by computer. If the trees were stratified into dis-crete layers, the area occupied by crowns would

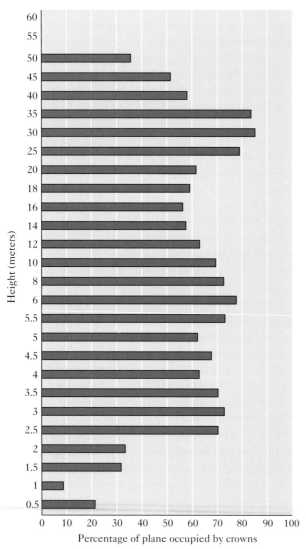

Vague peaks in the area occupied by crowns at different heights hint at the possibility of organized strata in a mature floodplain forest in Peru.

be expected to show peaks and troughs at different levels above the ground. Indeed, the results do suggest a series of peaks, but they are

not sufficiently pronounced to be statistically resolvable.

Nevertheless, it would be premature to dismiss the stratification hypothesis, because the data contain two kinds of noise that could blur any evidence of stratification. First, saplings must grow up through the spaces between strata on their way to higher positions in the stand. Second, the results represent average values for a whole hectare, and virtually no hectare in a tropical forest is spatially uniform. As noted in Chapter 4, tree falls create a patchwork within any mature forest. Since any microsuccessional patches within the hectare would not have attained structural equilibrium, their inclusion in the data would contribute to a perception of chaos. Further refinements of the analysis will be required to remove these sources of statistical noise.

While our frontal assault on the stratification problem is still in progress, we have taken another approach that answers a simpler question. We merely asked how many crowns were superimposed above the crown of each tree in the sample plots. Not surprisingly, we found that the lower the tree, the more crowns were growing above it. The average 20-meter tree was overtopped by 1.2 crowns, whereas the typical 7-meter tree languished in the shade of 2.0 taller crowns. From the analysis of smaller subplots, we determined that a hypothetical needle poked through the forest would encounter at least two additional crowns below the 7-meter tree. Random points on the ground in this forest are thus overtopped by an average of about five superimposed crowns, as Richards maintained. Our conclusion does not bear on the issue of stratification, however, since there is nothing in the results to suggest a regular spacing of the crowns in the vertical plane.

Richards's thesis also suggested that crown shape tended to vary systematically with vertical position. Emergent A-story trees were held to possess crowns that are broader than deep, while those of the B-story immediately below were reputed to be deeper than broad. To test Richards's claims, we computed an average shape factor for the crowns of each of the more common species of trees in our plots. Indeed, as Richards had maintained, the crowns of emergents were nearly twice as broad as deep, while those of species that mature at heights of 15 to 25 meters and never become emergents tended to have more equal breadths and depths. Species that mature below 15 meters again showed a tendency to have crowns that were broader than deep.

Once more our results seem to vindicate Richards's views. The systematic variation in crown shapes is compelling evidence that trees are fundamentally adapted for particular heights in the forest. Although these results do not constitute proof of stratification, they are consistent with the idea. We could expect that all trees that matured at a particular level would experience similar light conditions, no matter what the species of tree. If so, there would be a crown of a certain optimal shape that would allow foliage to receive the most light. In a process termed convergent evolution, natural selection might then mold the crowns of unrelated species so that they more closely approximated the optimal shape. Each species would then be adapted to a particular zone within the vertical reach of the forest. A tendency for these zones to cluster would result in stratification; if the

optimum zones for different species were staggered at various heights, there would be no stratification. The issue is still unresolved.

If the shapes of crowns do indeed vary systematically with vertical position in the forest, it would suggest that the light field may change in predictable ways from the top of the canopy downward. A-story emergents that rise above the general canopy receive full exposure to the sun and their crowns are free to expand laterally. Such trees can be expected to have broad, flat crowns, as already explained. The spread of some emergents can be truly dramatic. *Ceiba pentandra* is a prime example: the largest specimens, with spreads in excess of 40 meters, would cover nearly half a football field. One wonders how a tree can support such a massive crown without becoming precariously top-heavy.

The space underneath such a giant might be occupied by a dozen or more "B-story" trees. Here they would be permanently cut off from overhead illumination. Direct light would be available to them only during hours when the sun was low on the horizon. To absorb such low-angle light effectively, trees growing beneath emergents would need relatively deep and narrow crowns. In this respect they would resemble the conifers of the boreal forest that must also optimize reception of low-angle light.

Below the second tier of trees, in what Richards has termed the C-story, the light environment should be further transformed. Here, lighting conditions are even more complex. Below the gaps between A-story emergents, illumination could be expected to pass at high angles between the crowns of B-story species. Beneath emergents, the light will be filtered

Shrouded in the mist of dawn, a giant kapok *(Ceiba pentandra)* spreads massive limbs over dozens of lesser trees in this Peruvian floodplain forest.

through the crowns of both A- and B-story trees. Consequently, it will be broken up into a large number of small beams that are likely to penetrate over a wide range of angles.

From these simple arguments, it is easy to appreciate that the light conditions in a tropical forest will become more variable and complex as light passes through successive strata. The diversity of lighting conditions should be highest at mid-level, where a tree can experience any exposure to illumination from open sun to deep shade. An extremely irregular mosaic of conditions is created by the variable numbers of crowns overhead and the frequent small to large openings created by tree falls and broken branches. Therefore, it is perhaps not surprising that diversity among tree species is greater at mid-levels in the forest than in the upper and lower strata. Whether or not an idealized form of stratification exists is a question that still begs for resolution. Nevertheless, it is clear that the vertical dimension of tropical forests contains a complex gradient of light conditions, and that, as Richards proposed, species are adapted to occupy specific positions in this gradient.

Tree Diversity in Wet and Dry Tropical Forests

It should not be surprising that diversity is higher in a forest where as many as five species can spread their crowns over a single point on the ground. The superposition of many crowns is yet another mechanism that contributes to the high diversity of plant species in tropical evergreen forests. We can consider the stacking of plants one above another to be the vertical dimension of plant species diversity, to distinguish it from the horizontal component of diversity considered in the preceding chapter. The vertical dimension is able to accommodate high levels of diversity only where the climate and soil conditions permit trees to attain large stature. According to the sunfleck model, a second tree stratum can form only at a certain distance below the canopy, a rule that may apply to a third and successive strata as well. Only very tall forests provide the internal space to accommodate five strata.

We may draw upon this principle in seeking to explain why the species diversity of tropical dry forests is typically so much less than that of evergreen forests. Dry forests are characteristically of low stature and lack giant emergents. The canopy of many dry forests is only 15 to 20 meters in height, and trees taller than 30 meters are rare. Consequently we would expect dry forests to contain fewer superimposed tree strata than wet forests. Curiously, no one has looked to see whether this is true.

Some indirect evidence obtained by Alwyn Gentry seems to support this possibility, how-

A variegated squirrel *(Sciurus variegatoides)* opening a tamarind fruit in a Costa Rican dry forest. Squirrels of many species inhabit the world's tropical forests, where they vie with parrots for the seeds that form the mainstay of their diets.

ever. Gentry recorded the numbers of species of trees (greater than 10 centimeters in diameter at breast height) in 0.1-hectare plots from a wide range of sites in Central and South America. Plotting the number of tree species against the annual rainfall received at the sites, Gentry discovered that tree diversity in the plots increased strongly and linearly with rainfall from 500 mm to about 5000 mm, and did not increase further above 5000 mm.

Suspended from a high limb in a Costa Rican forest, an intrepid investigator looks down upon the crowns of mid-story tree strata.

Surprisingly, diversity in the canopy stratum of dry forests was no less than in the canopy stratum of wet forests. Instead, Gentry found that, as rainfall increased, virtually the entire increase in tree species diversity was contributed by the "subcanopy." The height of forest canopies also increases with rainfall over much the same range, although again, no one has thought to document the trend. Progressively taller canopy trees would allow more and more "subcanopy" species to take their places in the internal strata. Whereas a dry forest might contain trees greater than 10 centimeters in diameter in only one or two strata, three or four such strata might occur in a wet forest. Much, if not most, of the additional diversity of wet forests might therefore be contributed by increased packing of species in the vertical dimension.

Vertical Organization and Diversity

Now that we have gained some appreciation of the contribution made by the vertical dimension of tropical tree diversity, we are ready to return to the major question posed by this chapter and the previous one: Why do tropical forests contain so many more tree species than temperate forests? Perhaps now it will be possible to provide some definite answers.

Plant diversity is distributed in strikingly different fashions in temperate and tropical forests. At the latitude of Philadelphia, a forest might contain several dozen herbaceous species forming a ground layer, perhaps a dozen to twenty canopy tree species, a dozen to two dozen vines and shrubs, and one to three species in the mid-story tree "guild." The diversity is concentrated in the upper and lower strata; very few species occur in the middle. The dogwoods are almost alone in their guild.

It is the other way around in the tropical forest. There is always a herbaceous ground layer, but it tends to be poorer in species than the herbaceous layer of temperate forests. On the other hand, the canopy is far richer than its temperate counterpart. If tropical forests contained only these two strata, they would not be conspicuously more diverse than temperate forests, because a large number of temperate herbs would be balanced by an excess of tropical canopy trees. It is in the middle that the two types of forest differ so dramatically. Most of the 300 species of trees that Gentry found at his Yanomono site live beneath the canopy. Whereas the middle layer of a temperate forest adds insignificantly to the total diversity of the community, the middle layers of a wet tropical forest contribute perhaps half to three-quarters of the "extra" species of the tropical forest (excluding vines and epiphytes). Of all the mechanisms we have considered, increased vertical partitioning of the light field clearly makes the largest contribution to understanding the temperate-tropical gradient in tree diversity.

Several factors contribute to the difference in diversity. Tropical forests are taller than temperate forests, and therefore offer more space for internal strata. More light for photosynthesis is available in the tropics, where the sun passes almost vertically overhead and no energy is lost to winter. The total amount of light available to a temperate forest over a growing season is only about half of that available to a forest at the equator. Another contributory factor is that plants can grow and reproduce at lower light levels in the understory of tropical forests because, in the nonseasonal environment, they do not have to store energy for building new foliage in the spring. All these differences favor the development of additional vertically superimposed strata in wet tropical forests.

In contrast with the uniform light field found under the canopy of temperate forests, the light that passes through the emergent stratum of a tropical forest is spatially heterogeneous, a feature that allows species with distinct light-capturing adaptations to coexist at each level. The extreme variability of light conditions under the canopy probably contributes to the high diversity of the middle strata. The mid-story "niche" is consequently "larger" than that of a temperate forest. Here we find an analogy to the broader size distribution of insect prey that allowed more species to coexist in some tropical guilds of insectivorous birds.

In the last chapter we reviewed evidence that a number of additional mechanisms may also contribute to tropical tree diversity— stronger negative distance effects, more complex microsuccession, greater heterogeneity of gaps, a larger "regeneration guild," a greater variety of seed predators and dispersers, and, finally, the possibility of tighter species packing within guilds. Tighter packing could result from a more positive balance between speciation and extinction in the tropics. It is this latter possibility that we shall consider in the next chapter, "The Evolution of Species Diversity."

6

The Evolution of Species Diversity

Our inquiry into the causes of tropical diversity has apparently led us around in a circle. When we wondered how tropical forests could support so many more fruit- and nectar-feeding birds than temperate forests, it seemed reasonable to attribute the diversity of birds to the diversity of fruiting and flowering plants, and to a climate that permits plants to produce fruits and nectar throughout the year. Only a highly diverse forest can produce the variety of resources necessary to support a multiplicity of coexisting frugivores. Later, when we attempted to understand some of the factors that contribute to this very high plant diversity, we concluded that part of the answer lay

A member of the ancient edentate (toothless) lineage of mammals, long isolated on the island continent of South America, this tamandua (*Tamandua mexicana*) searches for ant and termite nests in the canopy of a Costa Rican forest.

in the variety and intricacy of the interactions between plants and animals—in particular between plants and the myriad fruit dispersers, seed predators and herbivores that inhabit most tropical forests. So many tree species were able to coexist in close proximity because the activities of animals and fungal pathogens prevented any one species from becoming too dominant. If this interpretation is correct, then some part of the plant diversity may be directly attributable to the diversity of animal species in the forest. But how could animal diversity contribute to plant diversity if plant diversity had to exist already in order to contribute to animal diversity?

To break out of this conundrum, we must ask how all the species of plants and animals came to be there in the first place. This same question arose earlier when we found that tropical guilds of birds contained many more species than their temperate equivalents. The "extra" guild members were attributed to tighter packing of species into these guilds, a likely consequence of a more vigorous proliferation of species through time. The myriad species that enrich the tropical world today evolved in the past, so it is logical to assume that the past may hold the key to the present. To understand how species might have accumulated in larger numbers in some parts of the world than in others, we must consider what may be termed evolutionary dynamics.

Evolutionary Dynamics

By evolutionary dynamics I refer to the fact that species evolve and go extinct, and that over millions of years the processes of speciation and extinction remain in approximate balance. There is no general tendency, in other words, for diversity to increase over geological time. Instead, the fossil record suggests that the diversity of mammals, dinosaurs, marine invertebrates, or almost any group of organisms is periodically dashed by major extinction events. Whether or not these extinction crises were caused by meteorite impacts, as some recent evidence suggests, numbers of them have punctuated the history of life on earth.

Diversity has recovered after each of these cataclysms, as the surviving organisms have given rise to new genera, families, and even orders, in a process known as adaptive radiation. After extinction crises, opportunities for new species to evolve are legion, as the sharply reduced species diversity relaxes competition and leaves some resources open to evolutionary opportunism (imagine, for example, the extinction of predators but not of their prey). After varying periods of time, but commonly 5 to 10 million years in rapidly speciating groups, the number of taxonomic units tends to return to approximately the same level as before the crisis, whereupon the numbers of species, genera, and families remain more or less stable for millions of years until the next mass extinction. These radiations often result in the replacement of major taxonomic groups by newly evolved functional equivalents. Examples are the replacement of condylarths from the Eocene Epoch by modern ungulates having more specialized feet and dentition, and the replacement of small-brained mammals known as multituberculates, which had relatively unspecialized teeth, by larger-brained rodents having gnawing incisors.

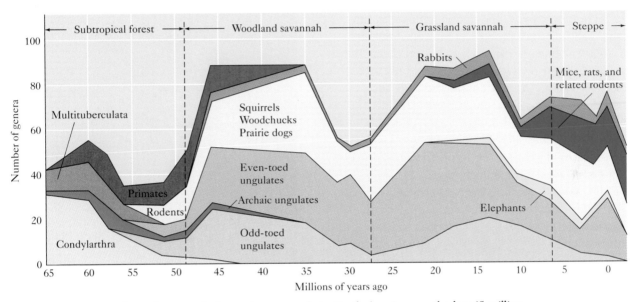

Numbers of genera of plant-eating mammals in North America over the last 65 million years. The record shows several extinction crises. Following the crises diversity recovered rapidly and thereafter the number of genera remained relatively stable. No systematic trend toward increasing or decreasing numbers of genera is evident.

The study of evolutionary dynamics, as deduced from the fossil record, leads to some important conclusions. There are periods when speciation exceeds extinction, periods (usually brief) when extinction exceeds speciation, and periods when the two remain in approximate balance. It is significant that the periods of balance occur at high, not low, levels of diversity, suggesting the existence of feedback mechanisms that limit diversity. The simplest sort of feedback is demographic. The more species there are to compete for a given supply of resources, whether the resource be fruit, prey, or solar energy, the lower the mean density of each species. If new species continue to evolve, at some point others will become so rare that they go extinct, and a balance will be achieved.

Throughout the history of life, whether on land or in the sea, the warm, nonseasonal equatorial regions have exhibited higher levels of diversity than the cooler, more seasonal environments of the high latitudes. High species diversity is thus intrinsic to the tropical belt, independent of geological period, the momentary disposition of drifting landmasses, or variation in global temperature or precipitation. Any comprehensive theory of species diversity must accommodate these undisputed features of the fossil record. To delve further into the issue of how evolutionary processes contribute to patterns of species diversity, we must review some of the basic features of the process of speciation, the evolutionary mechanism by which new species arise.

The term *speciation* comprehends two fundamentally distinct processes: the transformation of coherent lineages through time and the bifurcation of lineages. Despite the title of his revolutionary work, Darwin considered only the first of these processes; the second remained a mystery for nearly a hundred years after publication of *On the Origin of Species*.

It is obvious that, by itself, the transformation of lineages through time does not increase the diversity of species. Instead, species give rise, either slowly or rapidly depending on the circumstances, to descendent species, termed chronospecies. The ancestral forms then cease to exist. The total number of species remains the same, or decreases whenever there are extinctions.

It is only the second type of speciation, the bifurcation of lineages, that generates diversity. In this process, an ancestral species may give rise to one or a series of descendent "daughter" species, but does not necessarily go extinct in doing so. The total number of species may consequently increase. The study of this process is a major theme in evolutionary biology. It is an enormously varied and complex topic, one that has been the subject of many learned tomes. In a work such as this, we must necessarily make short of it, emphasizing only the main points.

The basic issues are how, how quickly, and under what circumstances do daughter lineages form and diverge in their biological properties from the parent lineage. When a daughter species arises in the presence of the parent species, the process is termed "sympatric speciation." When daughter species arise in geographically separate locations, so that the di-

A partner in a highly specific mutualism, this bee (*Eulaema* sp.) visits an orchid in the Peruvian rain forest. The pollen is precisely placed on the bee's body in a way that helps ensure that only individuals of the same species are pollinated. Male *Eulaema* bees are attracted from a great distance to this particular orchid by volatile aromatic substances. The bees, in turn, collect the attractants and later use them to entice females during courtship. The attractants and the placement of pollen on the bee reduce the likelihood of hybridization.

verging components of the parent population are in genetic isolation, it is termed "allopatric speciation."

Sympatric speciation can occur in at least two ways. New species of plants occasionally arise when individuals of the same species are able to fertilize themselves and double the complement of genes. Hybridization of existing species can also result in sympatric speciation in a process thought to be uncommon in plants and rare in animals. Otherwise, biologists have had a great deal of difficulty imagining how, in the absence of self-fertilization and hybridization, daughter lineages could diverge while the respective populations were able to interbreed. Even low levels of interbreeding lead to homogenization of the two gene pools, a process that is incompatible with genetic divergence. In general, sympatric speciation appears to represent occasional exceptions in a broad pattern of allopatric speciation, especially in such groups as birds and rain forest trees, which possess elaborate mechanisms that effectively inhibit hybridization.

The Allopatric Model of Speciation

In the allopatric model of speciation, populations undergo divergence both in genetic makeup and in physical form and structure while they are in isolation from one another. The diverging populations are normally separated by distance and by so-called barriers to dispersal. In its most basic form, the allopatric speciation process consists of two steps. First, some event must bring about the geographical isolation of two or more components of an initially contiguous and interbreeding population. Second, once geographical isolation has been achieved, one or both of the daughter populations must evolve new traits. These two steps may be sufficient to increase the total number of species in the radiating group, but they are not sufficient to increase diversity at the local level. This requires a third step, the establishment of sympatry. One or both of the daughter species must extend its geographic range so that the two come into contact. If the two species are now different enough in reproductive behavior not to interbreed and sufficiently distinct ecologically, they may come to occupy the same habitat. Only when all these conditions are met can two daughter species achieve complete coexistence and maintain their genetic integrities.

This abbreviated account leaves many questions unanswered. One of the most crucial is how geographic isolation can occur in a reversible fashion, so that populations are initially split into separated components and then later reunited. Another unanswered question is how long daughter populations must remain apart to achieve the genetic and ecological differences necessary to permit coexistence without interbreeding. Various lines of evidence suggest that the populations must be separated for a long time, measured in tens or hundreds of thousands of years. We are then compelled to imagine how populations can be split and subsequently rejoined after intervals that are neither too long nor too short to be consistent with the fossil record.

Subpopulations can become geographically isolated in either of two ways. So-called long-distance dispersal can occur when a seed, pregnant female, or small group of individuals travels or is passively carried to a new location. For example, seeds may be transported on the foot of a migrating bird. Long-distance dispersal is the only means through which oceanic islands

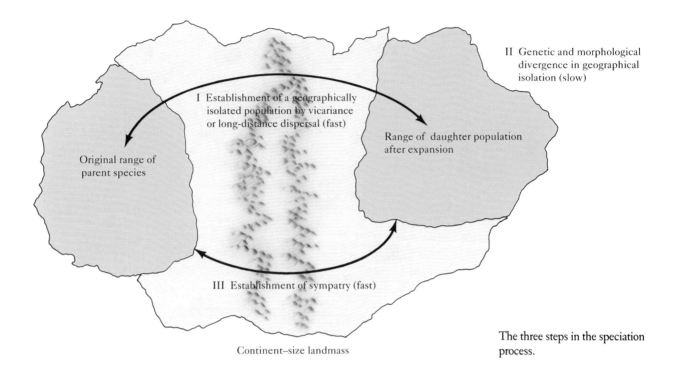

II Genetic and morphological divergence in geographical isolation (slow)

I Establishment of a geographically isolated population by vicariance or long-distance dispersal (fast)

Range of daughter population after expansion

Original range of parent species

III Establishment of sympatry (fast)

Continent–size landmass

The three steps in the speciation process.

are colonized by plants and animals. Presumably, a population may establish a beachhead on a mainland through long-distance dispersal, but supporting evidence is scanty.

Geographically isolated populations can also arise through a process termed *vicariance.* Vicariance occurs when a population is split by natural geological or meteorological forces. Unlike long-distance dispersal, vicariance is always a passive process from the point of view of the organism. The intervention of lava flows or mountain ranges, changes in sea level, and the drifting apart of continents are some of the more dramatic mechanisms of vicariance. All of these events can and do occur in geological time, but most seem to lack the required feature of reversibility. The splitting of populations by vicariance may eventually lead to an increase in global diversity, but not to the local diversity that is our principal concern.

Before continental drift became a scientifically acceptable concept in the 1960s, biogeographers indulged in amazing mental exertions to explain the distribution of living organisms solely through long-distance dispersal. All sorts of improbabilities were contemplated, including the rafting of seeds and eggs across the Pacific ocean in tree trunks. Now the fashion has reversed, and vicariance is being touted as the principal mechanism of geographical isolation.

So-called vicariance biogeography began to gain adherents after geologists had become convinced, as the spurned German meteorologist Alfred Wegener had advocated some 50 years before, that the continents really do change their relative positions, with accompanying fission and fusion of the principal landmasses. By reconstructing patterns of continental movement and extrapolating them backward in time, geologists affirmed the existence of a great southern continent, Gondwanaland. To biologists this was a most welcome revelation, because it provided an explanation of some otherwise embarrassingly awkward distribution patterns, such as the occurrence of boas and iguanid lizards in Madagascar, Fiji, and the New World tropics and the occurrence of *Nothofagus* and *Araucaria* forests in Australia, New Zealand, and Chile. Still, continental drift did not seem to provide the requisite conditions for increasing local species diversity, except in the rare instances in which long-isolated landmasses happened to collide.

Just such an event took place 3.5 million years ago when colliding crustal plates uplifted a loose archipelago of islands to form the Panamanian isthmus. Before this event, North and South America had experienced independent evolutionary histories for about 100 million years.

The Great American Faunal Interchange

The formation of the Panamanian land bridge constitutes one of the great natural experiments of all time. By carefully analyzing the fossil record from just before and just after the joining of the continents, paleontologists have been able to determine whether the merging of two totally distinct faunas would lead to an increase in diversity. Both the North and South American continents carried large and diverse assemblages of organisms. The mammals, in particular, were strikingly dissimilar.

South America had been an island continent for some 80 million years, since breaking away from Africa in the age of dinosaurs. Archaic groups accordingly predominated in its mammal fauna, including such bizarre and now extinct creatures as glyptodonts (giant tortoise-like armadillos), toxodonts (a group of hooved ungulates), and ground sloths. Like Australia,

A giant ground sloth, representative of the pre-isthmian fauna of South America. After the formation of the Panamanian land bridge, ground sloths of several sizes became widespread in both North and South America, persisting until a few thousand years ago when they were hunted to extinction by native Americans.

which had been similarly isolated, South America contained a wide spectrum of marsupials, not only opossums of many types, but other

A three-toed sloth (*Bradypus variegatus*) naps in a Panamanian *Cecropia* tree. Sloths earn their name from their languid and indolent habits, which are actually energy-conserving adaptations to a low-calorie diet of leaves.

more formidable beasts, led by *Thylacosmilus,* the marsupial saber-toothed tiger. Meanwhile, North America had been periodically in contact with Eurasia via the Alaska land bridge and carried a more familiar fauna consisting of wild cattle, horses, cats, dogs, deer, squirrels, camels, tapirs, elephants, and many others.

Before what is known as the "Great American Interchange," North and South America each harbored 30 to 35 families of mammals, most of them distinct. The potential thus existed for the diversity of mammals to double at a single stroke with the closing of the isthmus. However, this is not what happened. Today, North America supports 33 mammal families, and South America 35. Families of North American origin contributed 40 percent of the families in contemporary South America, and South American families constitute 36 percent of contemporary North America's mammal fam-

ilies. There has indeed been a substantial interchange, but it has not increased diversity at the family level because extinctions have matched the flow of alien lineages across the isthmus.

At the generic level, the details are somewhat more complicated and less readily interpreted. Before the formation of the land bridge, the number of mammal genera in South America had hovered around 72. With the appearance of North American forms, generic diversity rose rapidly to over 100, as the immigrants added to the native fauna. In North America, however, quite the opposite occurred. The pre-landbridge fauna contained 131 genera, but this number actually dropped to 101, perhaps for unrelated reasons, as South American forms began to appear. Subsequently, the number of known genera in both continents has risen substantially as immigrant and native lineages have diversified. North America now contains 141 genera (compared to 131 before) and South America 170 (formerly 72).

The large increase in the diversity of South American genera is not to be taken literally. Comparisons of past and present numbers are misleading because the current number represents a comprehensive count of living forms, whereas the pre-isthmian total represents the fossil forms known from a limited number of localities, most of them in subtropical or temperate Bolivia and Argentina. The pre-isthmian mammals of the tropical portion of the continent, by far the larger part, are almost completely unknown, and their number is likely to have been significantly greater than has been documented in the fossil record.

Putting aside this probable artifact of the accidents of preservation, evidence provided by the Great American Interchange strongly supports the notion of an equilibrium between the rates of appearance of new forms (here through immigration rather than speciation) and extinction of old forms. Appearances and disappearances of taxa tend not to be tightly coupled in time, making it difficult if not impossible to pinpoint cause and effect in particular cases. But they do seem to balance out over millions of years to maintain diversity levels (of genera) within approximately a factor of two. Considering the complexity of the processes involved, this seems remarkably tight control.

The separation of South America from Africa initiated a cycle of vicariance that ended after some 80 million years with the formation of the Panamanian isthmus. Cycles of vicariance of that length are not particularly relevant to understanding species diversity at either the local or the continental scale. The fossil record suggests that considerably shorter cycle times

are required to account for the rapid recovery of diversity after periodic extinction crises. The ensuing radiations of new forms are often complete in 5 to 10 million years. For such rapid speciation to occur, vicariance cycles would have to have much shorter periods, in the range of 1 million years or less.

The fact that the cycles may be neither too short nor too long puts severe constraints on the types of vicariance mechanisms that can generate diversity. Geological events such as continental drift and mountain building normally take too long, as do the associated changes in sea level. Short-term climatic fluctuations, such as those driven by the sunspot cycle and major volcanic eruptions, are far too short.

There is, however, a class of perturbations that lies squarely within the required range of time values. These are the so-called Milankovitch cycles, oscillations in the earth's orbital parameters, named in honor of the Czech mathematician who first described them in 1930. The shape of the earth's orbit varies from more round to more elliptical in cycles of 413,000 and 100,000 years. These two cycles result from the gravitational interactions of the earth with the other planets. When the planets are well dispersed, the forces are relatively symmetrical and the earth's orbit assumes a more circular form; when several planets become aligned on one side of the sun, their combined pull significantly distorts the earth's orbit. In addition, Milankovitch described two other cycles: (1) the tilt of the axis of rotation increases and decreases with a 41,000-year periodicity, and (2) the season of the year during which the earth is closest to the sun gradually shifts from summer to winter and back on a 22,000-year cycle. The four Milankovitch cycles interact in

North and South American mammals migrated in two directions after the closing of the Panamanian isthmus. South American forms that migrated northward are shown in the upper part of the diagram, and North American forms that migrated southward in the lower part.

a complex way to generate significant changes in the earth's climate over periods of a few tens of thousands to hundreds of thousands of years. It is possible that these cycles constitute the principal engine of speciation on earth.

Milankovitch Cycles and a Possible Speciation Pump

Geologists now generally acknowledge that Milankovitch cycles were responsible for the numerous advances and retreats of continental glaciation during the Pleistocene Epoch, the geological era covering the last 1.5 million years. While glaciers were waxing and waning in the Northern Hemisphere, the climate was also changing elsewhere in the world, including in the tropical belt.

One of the first biologists to appreciate the significance of such changes was the British ornithologist R. E. Moreau, who proposed that the central African forest had expanded and contracted during the Pleistocene. Moreau believed that the reversible fragmentation of African forests provided the vicariance mechanism needed to explain the distribution of endemic (geographically restricted) species in many now isolated East African forests. Moreau failed to support his arguments with persuasive evidence from geology and meteorology, however, and was largely ignored by a skeptical scientific public.

The next such proposal to appear was far better documented. Its author is Jurgen Haffer, a world-traveled German petroleum geologist.

Haffer is a remarkable renaissance man, a person who has distinguished himself in two careers, one remunerative, the other a labor of love. For several years during the 1960s, he was stationed with an exploration team in northern Colombia analyzing geological structures. He thus acquired an authoritative knowledge of the recent geological history of the South American continent. At the same time, he was a bird-watcher who devoted his off hours to studying Colombia's incomparably diverse avifauna. In 1969, he merged his vocation and his avocation in a landmark paper that appeared in the prestigious journal *Science*. It is fair to say that this paper revolutionized our thinking about speciation and about the role of vicariance in particular.

Haffer's thesis begins with the observation that the contemporary distribution of rainfall over tropical South America has given rise to both forested and nonforested habitats. The forests are primarily in Amazonia, whereas grassland, savannah, and savannah woodland compose the llanos of Colombia and Venezuela and the cerrado and campo cerrado of Brazil. Haffer observed that the limits of these nonforest habitats closely coincide with an annual precipitation of 1500 millimeters (60 inches). Where annual rainfall is above this level, the vegetation is predominately forest, and where it is less, dry-season fires convert the landscape to grassland or savannah.

Haffer then proposed that during the Pleistocene, as glaciation advanced over the northern continents, rainfall in the tropical regions was reduced. Lower precipitation would be expected during periods of glaciation because a general lowering of the earth's mean

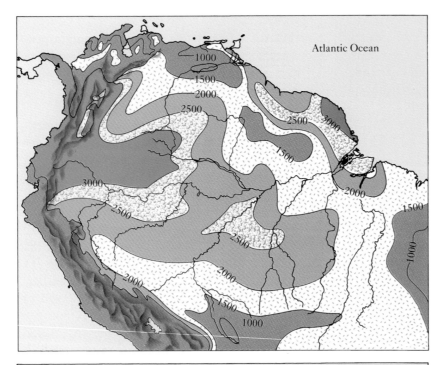

The contemporary pattern of rainfall in South America. Areas receiving more than 2500 millimeters of rainfall annually are postulated as the locations of Pleistocene forest refugia. Savannahs and other types of nonforest vegetation occur in areas receiving less than 1500 millimeters.

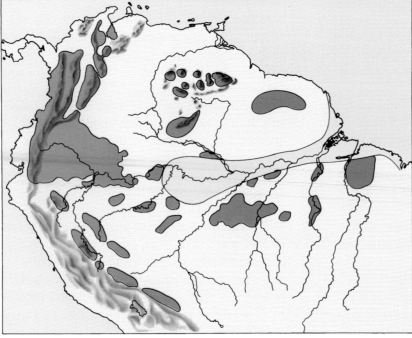

Areas proposed by Haffer as Pleistocene refugia for forest-dwelling plants and animals in South America. Populations isolated for tens of thousands of years in such refugia often evolved distinctive characteristics, such as the distinct forms of the butterflies on page 145.

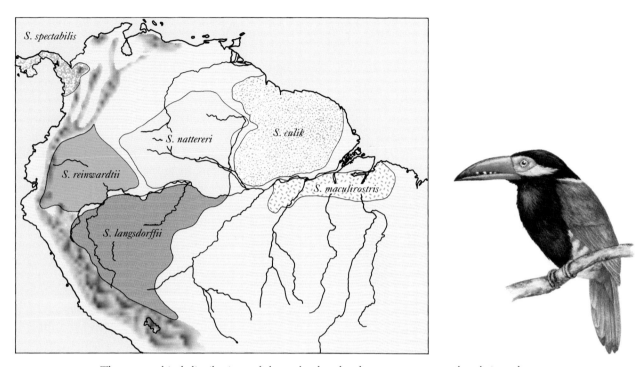

The geographical distributions of these closely related toucanets suggest that their evolutionary origins lie in the Pleistocene refugia depicted in the map on page 142. Although neighboring species often have abutting ranges, hybrids between the forms are rare or nonexistent, indicating that reproductive isolation is complete. No two of these toucanets have yet achieved ecological compatibility, however.

temperature would reduce evaporation. The record of fossil pollen from Andean lakes indicates that mean temperatures in South America fell periodically by 4 to 6°C. The distribution of rainfall in South America is such that a modest decrease of only 20 to 25 percent would greatly expand the area receiving less than 1500 millimeters per year. Haffer offered a hypothetical map showing the distribution of forest and nonforest habitats under a 25-percent reduction in rainfall. It shows the forest to be broken into a number of separate fragments, or "refugia."

We now have a perfect cyclical vicariance mechanism, referred to by some as a "specia-

tion pump." During dry periods, forest-dwelling species would be confined to widely scattered refugia. When wetter conditions returned, the various pockets of forest would expand and eventually coalesce into the vast contiguous forest that currently covers most of the Amazon basin. If the dry periods were to correspond roughly to the periods of glacial advance in the Northern Hemisphere, they would have lasted 50 to 100 thousand years or more, possibly sufficient time to permit the divergence of isolated sister populations.

As evidence that such evolution had indeed occurred, Haffer offered maps showing the distributions of closely related but distinct forms of toucans, chachalacas, and other birds. The distributions of these birds coincided strikingly with the regions of the postulated refugia. Haffer also looked for zones of hybridization in areas between refugia, where incompletely differentiated descendent lineages should have spread out with the advancing forest upon the return of moist conditions. He was able to document the occurrence of several hybridization zones just where they would be expected to lie.

The publication of Haffer's paper unleashed a flurry of research activity. So many biologists engaged themselves in the examination of his ideas that an international symposium was called in 1978 to evaluate what had become known as the "refuge hypothesis." Dozens of papers in the hefty volume that ensued enthusiastically, if sometimes uncritically, endorsed Haffer's thesis. Distribution patterns in group after group of organisms were reported to reflect the former existence of forest refugia. Dissenting voices were few, the most notable that of John Endler, now professor of biology at the University of California at Santa Barbara.

Endler argued that Haffer's evidence could be explained by other means. By way of counterargument, he produced an analysis of putative refugia in tropical Africa, showing that they coincided with areas of environmental uniformity. New species might originate in large, relatively homogeneous areas of habitat, Endler suggested, in which populations evolved particular adaptations in accord with the characteristics of the local environments. If the areas postulated as refugia in South America also tended to be relatively homogeneous within themselves,

but distinct from one another, then divergence might occur *in situ* even without the physical fragmentation of the habitat. If this were true, then zones of intergradation or hybridization would be expected to fall in regions of maximal environmental change lying between the areas of uniformity. In the absence of undeniable physical evidence of the former existence of savannah habitats in areas now occupied by Amazonian forest, it is difficult to give either Haffer or Endler the last word.

More recently, new voices have been added to the relatively faint chorus of dissenters. One is that of Paul Colinvaux of Ohio State University. Colinvaux has analyzed pollen deposited in the sediments of lakes at the foot of the Ecuadorean Andes. Pollen grains reveal information about the temperature at the time of deposition, because the grains can be identified to genus or even species. For example, Colinvaux found pollen of *Podocarpus*, a relative of pines and firs, in the sediments of lakes in the Andean foothills at an elevation of 1100 meters. Today, *Podocarpus* occurs in the Andes only at elevations at least 700 meters higher, where the mean annual temperature is 4.5°C (8°F) lower than at 1100 meters. If such depressed temperatures extended across the entire Amazon basin, the impact on vegetation must have been appreciable. However, no fossil pollen has yet been found to support the presence of savannah grasslands in areas now occupied by forest, perhaps only because of an extreme paucity in Amazonia of sites offering long-term pollen records. This second challenge to the refuge hypothesis must also be regarded as a standoff.

Populations of the butterfly *Heliconius erato* are found in geographically distinct areas that suggest the locations of Pleistocene forest refugia in South America. Although some populations are strikingly distinct in appearance, all these forms are considered to belong to the same species.

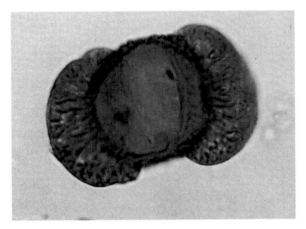

A fossil pollen grain of *Podocarpus*, a high Andean relative of northern conifers.

With the advent of ever more sophisticated means for assessing genetic relationships, a third argument has come to the forefront. Using a technique known as DNA hybridization, molecular biologists can measure the amount of genetic divergence to have occurred in any two evolutionary lineages since they shared a common ancestry. In general, the genes of closely related species possess many more DNA sequences in common than do the genes of more distantly related forms. If it is assumed that genetic change progresses at a constant rate, differences in DNA sequences can be used to estimate the time since a particular organism began to diverge from related forms. DNA can thus provide a "molecular clock."

To measure something, it is necessary to have a calibrated measuring stick. Molecular biologists have calibrated the molecular clock by comparing levels of genetic divergence in lineages that were separated at well-dated times in the past, such as the closing of the Panamanian isthmus. From such comparisons, investi-

gators determine a standard rate of genetic change, which they then apply to the lineages under study. A potential weakness in this procedure is that the rates of genetic divergence may not be equal in unrelated lineages; some may evolve rapidly and others slowly. If the clock is calibrated with one group of organisms, it may give erroneous estimates of divergence times if genetic change proceeds more or less rapidly in another lineage. Nevertheless, it has become possible in principle to test Haffer's hypothesis by estimating the dates of divergence of forms that putatively evolved in Pleistocene refugia.

Joel Craycraft and Richard Prum are two ornithologists who have applied molecular biology to the study of evolutionary relationships among birds, and they have recently begun to tackle this very problem. Using toucans, one of Haffer's prime examples, as a test case, Craycraft and Prum deduced that the existing species arose in a variety of situations at various times in the past. For example, related lineages now occupying Central America and the Amazonian region seem to have been split by the Andean uplift, whereas similar forms inhabiting different portions of Amazonia must have originated under other circumstances. At present, however, molecular clock technology is not sufficiently refined to confirm or reject the hypothesis that modern toucan species originated during the Pleistocene as Haffer supposed. If subsequent research were to support an earlier time of divergence, then the whole notion that Pleistocene climatic oscillations drove a "speciation pump" may have to be abandoned. For the time being, however, Haffer's ideas have not lost their luster.

The idea of a flurry of Pleistocene originations has enormous appeal, not only because it provides a plausible mechanism of cyclical vicariance, but also because it can partly explain tropical diversity. The maps prepared by Haffer or any of his many followers show numerous candidate refugia in tropical South America. A single vicariant event such as the drying of the continent may have served to isolate as many as a dozen sister lineages. By the time wetter conditions returned several tens of thousands of years later, the continent might have been enriched by scores of new species. Repeat the process a few times (there were at least four major glacial cycles in the Pleistocene), and it would not be difficult to account for the 3000 bird species that now inhabit the continent.

This scenario can be contrasted with the better-documented one that unfolded contemporaneously in North America. When glaciers moved southward, covering nearly the entire landmass down to New Jersey, Illinois, and Kansas, the familiar habitats of that land virtually ceased to exist. Boreal forest ran through Virginia and Tennessee, and pollen traces of what we would recognize as the eastern deciduous forest have been found only along the Gulf Coast. Under these dramatically altered conditions, the plants and animals that frequent the central and southern states had only two possible refugia: the Florida Peninsula and northern Mexico.

The evidence is scanty that new species originated in North America during periods of glacial advance, and the proliferation of new forms would certainly have been subdued relative to that in tropical South America. Herein lies the almost irresistible appeal of the Haffer thesis. It offers a plausible mechanism of cyclic vicariance and at the same time promises to account for the elevated species diversity of tropical habitats. But even if it is eventually substantiated beyond reasonable doubt, the Pleistocene refuge hypothesis must be regarded as a special case so far as tropical diversity is concerned.

Tropical Diversity: The General Case

Tropical floras and faunas have always been more diverse than temperate ones, and therefore a general theory of tropical diversity cannot rely on cycles of glacial advance and retreat, for these have not been a regular feature of the earth's environment. In fact, the previous such episode took place in the Permian Period, some 300 million years ago. Otherwise, the earth has been appreciably warmer than it has been during the last two million years. Is there something else about the tropics that might favor the proliferation of species?

In fact there is. The tropics are simply larger, and in large landmasses the possibilities for vicariant isolation and reestablishment of contact improve substantially. We know this from an examination of the faunas of islands. Islands, even quite small ones, often shelter endemic species. The presence of an endemic species, however, does not imply that the species originated within the shores of the island. Instead, the endemic is more likely to be a

Unique to Madagascar, this fossa (*Cryptoprocta ferox*), a relative of civets and mongooses, is the top mammalian predator on that island.

relict or chronospecies derived from a now extinct progenitor on the mainland. Even when two species of the same genus are endemic to an island, close scrutiny has usually revealed that both of them are more closely related to other species living on nearby islands or the mainland than they are to each other, implying that the founding populations arrived independently. Only when islands harbor two or more related species in endemic genera or families can a strong case be made that allopatric speciation has taken place within the confines of the island.

If we now ask what is the smallest island from which there is convincing evidence of speciation in situ, we find that the answer varies greatly with the taxonomic group in question. For groups having relatively limited powers of dispersal, such as reptiles and amphibians, islands the size of Jamaica are apparently sufficient. For groups having moderate dispersal ability, such as higher plants and small mammals, more space is required, areas equivalent to those of New Caledonia or Cuba. And for rapidly dispersing birds, bats, and large mammals, only the world's largest islands, Madagascar and New Guinea, appear adequate.

These results provide indirect but persuasive evidence leading toward two conclusions: first, most, if not all, of the speciation in these groups is allopatric, and second, the allopatric

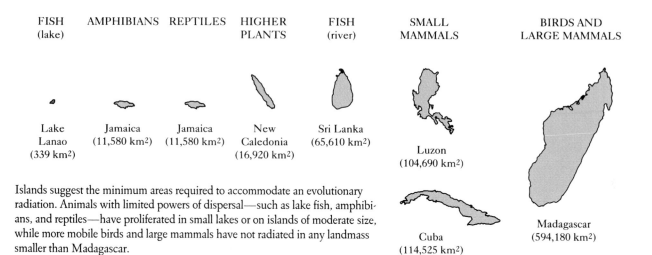

| FISH (lake) | AMPHIBIANS | REPTILES | HIGHER PLANTS | FISH (river) | SMALL MAMMALS | BIRDS AND LARGE MAMMALS |

Lake Lanao (339 km²) Jamaica (11,580 km²) Jamaica (11,580 km²) New Caledonia (16,920 km²) Sri Lanka (65,610 km²) Luzon (104,690 km²) Cuba (114,525 km²) Madagascar (594,180 km²)

Islands suggest the minimum areas required to accommodate an evolutionary radiation. Animals with limited powers of dispersal—such as lake fish, amphibians, and reptiles—have proliferated in small lakes or on islands of moderate size, while more mobile birds and large mammals have not radiated in any landmass smaller than Madagascar.

speciation mechanism can go to completion only in large and geographically complex areas. That is, only in such large areas will the allopatric speciation mechanism generate increased local diversity. We might further infer that the larger the area, the greater the rate of speciation.

Every species of plant or animal lives within a circumscribed range of environmental conditions. In any particular case, we can measure this range in terms of physical or biotic parameters such as temperature, elevation, rainfall, soil type, and habitat. For speciation to occur, two patches of suitable habitat must remain isolated long enough for the isolated lineages to diverge. Sufficient time is counted in tens or hundreds of thousands or perhaps even millions of years. The existence of Milankovitch cycles implies that climatic conditions are continuously, even if reversibly, changing. In such a kaleidoscopic world, only a large and topographically varied area is likely to contain a given type of habitat for an adequate time to allow diverging allopatric populations to become distinct species.

Long-persisting areas of adequate size are more likely to occur in the tropics for three complementary reasons. First, the earth is roughly spherical, which means that the area between successive degrees of latitude is highest at the equator. Second, climates are symmetric across the equator, so the tropics form one continuous belt, whereas equivalent temperate climates in the Northern and Southern hemispheres are separated by thousands of kilometers. And third, mean annual temperature varies little with latitude near the equator, so that a broad belt of almost constant temperature conditions extends out to 20 or more degrees of latitude. As a consequence of all three factors, warm, humid, relatively aseasonal climates form a virtually continuous belt from central Mexico to Bolivia, spanning 40 degrees of latitude. Scores of species of birds, mammals, and trees occur over this entire range. A temperate

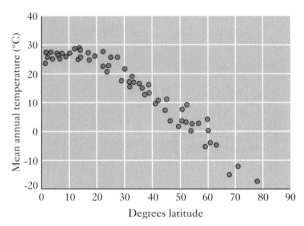

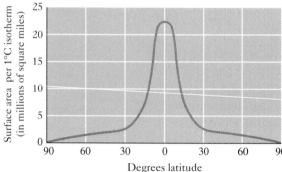

(Top) The annual mean temperature of low-elevation continental localities does not noticeably decrease from the equator out to a latitude of about 25 degrees. At higher latitudes, mean temperature decreases steadily to the poles. *(Bottom)* The area of the earth's surface between one degree Centigrade isotherms of mean annual temperature decreases as latitude increases. The area of habitat available to support the speciation process is more than 10 times greater at the equator than at mid latitudes.

different latitudes, one can simply compute the area available on the earth's surface between one-degree isotherms of mean annual temperature. The area between successive one-degree isotherms is roughly 11 times more extensive at the equator than at 40 degrees, the latitude of Philadelphia or Madrid. It should not be a concern that there is more land surface at mid-latitudes in the Northern Hemisphere than at the equator today, because the current disposition of the continents is an ephemeral condition. More important, the greater extent of tropical habitats is primarily the consequence of climatic symmetry and the flattening of the global temperature gradient near the equator, and only secondarily of the diminishing area between successive degrees of latitude.

Here we have a time-invariant feature of the earth's geography that predicts greater opportunities for allopatric speciation around the equator than at higher latitudes. In addition, extinction rates in the tropics should be lower because large areas tend to support large populations which are less vulnerable to extinction than small populations. The effect of geography on speciation and extinction is illustrated in the figure on the facing page.

The probability of extinction is considered to vary inversely with population size. Species confined to small areas are therefore vulnerable because the size of their populations is proportional to the available area of appropriate habitat. Because many temperate habitats are relatively small in area, extinction rates at temperate latitudes will climb more quickly as the diversity of species increases. Conversely, in the tropics, extensive areas of habitat such as the Amazonian rain forest accommodate high rates of speciation. The rate of extinction is less sen-

counterexample would be a species that lived everywhere from south Florida to the arctic circle. There are a few species that do, but very few, and they are mostly birds that escape the arctic winter by migrating.

To roughly model the amount of habitat of a given kind that is available to populations at

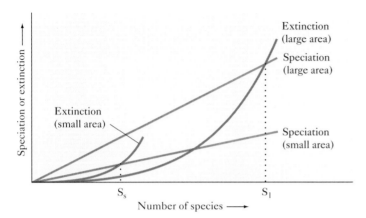

How speciation and extinction might vary with the number of species present in a small and large landmass. Speciation is assumed to depend on the number of species available to speciate and on the opportunities for geographical isolation within land areas of different size. Extinction is expected to increase with the number of species present—faster in a small area. The expected numbers of species at equilibrium, S_s (small area) and S_l (large area), are found where the rates of speciation and extinction are in balance.

sitive to increasing species diversity than in temperate regions because each species is able to occupy a larger area. Species diversities at equilibrium are higher in the tropics, and therefore packing levels are higher as well.

In conclusion, we find that despite the enthusiastic acceptance of the Pleistocene refuge hypothesis, it emerges at best as a special case, applicable only to the most recent phase of the earth's history. Fundamental features of the earth's climate and geography provide a more general explanation for tropical diversity, but this explanation leaves open the nature of the vicariance mechanisms that generate the species. Probably these are many and diverse, and include long-distance dispersal as well as geological events such as the uplifting of the Andes.

Nevertheless, one mechanism of vicariance is bound to be more important than any other. I suspect that Milankovitch cycles provide this mechanism. Just as the spherical shape of the earth and the symmetry of climates across the equator are intrinsic features of our planet, so are Milankovitch cycles inherent to our place in the solar system. They ensure that climate will

change, and that it will change reversibly over a period of time consistent with the rates of biological diversification observed in the fossil record. Although the climatic changes produced by the Milankovitch cycles are not large, only a small change is needed to cross one of the many thresholds in ecology that can transform one type of habitat into another. One such threshold is Haffer's example of the boundary between forest and savannah being set by a rainfall of 1500 millimeters. It can thus be appreciated that relatively small changes in climate can produce large changes in the distribution of habitats, and that the climatic changes generated by the Milankovitch cycles can and will be sufficient to function as vicariance mechanisms.

The larger area occupied by tropical environments, in conjunction with climatic symmetry across the equator, ensures that global climate change will generate more vicariant habitat islands in the tropics than at temperate latitudes. To this extent Haffer's idea is on the mark, but, as it is based on proximate rather than ultimate causes, it provides only part of the answer.

7

Convergence or Nonconvergence?

When we use the term *tropical rain forest* in everyday conversation, in news releases, and even in the scientific literature, it carries a ring of universality. For the typical resident of the temperate zone who has never seen a tropical forest, the term invokes the same mental image of dense green foliage and brightly colored birds, no matter what part of the world is under discussion.

Armed with a larger store of factual detail, scientists have tended to take a more particular view. Here is where a certain species can be found, there another. Yet, from a gestalt perspective, one forest tends to look rather like another. Wherever one goes, there are the same typically tropical plant forms—epiphytes, palms, heavy li-

A red howler monkey dines on fresh young foliage as this deciduous forest in Venezuela bursts into leaf at the end of a long dry season.

anas, stilt roots, flaring buttresses, and scores of plant species with indistinguishably similar elliptical leaves. It is thus not surprising that the prominence of these plant adaptations in tropical forests around the world has given strong support to the idea of evolutionary convergence—the theory that life forms in similar environments independently evolve similar adaptations.

Convergence is the ultimate paradigm of the evolutionary process. Natural selection winnows randomly occurring mutations to produce adaptations that confer superior performance in the context of a particular set of stresses and opportunities. The warm, moist, aseasonal climate of the humid tropics seemed the ideal proving ground for the convergence paradigm. The climate in Manila cannot be distinguished from that in Colombo or Belen. Morphological and physiological characteristics that proved successful in one part of the tropics should therefore succeed in others. Given identical conditions in which to operate, evolution should produce consistent results anywhere, if given enough time. Testing the proposition turns the whole world into an evolutionary laboratory.

Richards's ideas on the stratification of tropical forests arose out of comparisons of sites in the New World, Africa, and Asia. He found evidence that the forests of these regions have a similar vertical organization, and these similarities lent support to the notion of convergence. The visually appealing profile diagrams that Richards and other botanists have used to describe tropical forests clearly demonstrate that they are not all alike, but leave the impression that the differences are due to local effects of soil, topography, and drainage. Running

through the literature is the implicit assumption that perfectly matched sites would be structurally similar, regardless of whether they happened to be on the same continent.

Convergence in the Botanical Realm

Botanists have been particularly receptive to the idea of convergence in the rain forest, in part because the great antiquity of tropical forests should have allowed plants "enough time" to evolve similar adaptations. Flowering plants (angiosperms) first appeared in the fossil record about 150 million years ago during the age of dinosaurs. The angiosperms include nearly all familiar plants except pines and other conifers, and ferns. The evolutionary innovations that distinguish the angiosperms from the gymnosperms, seed ferns, and several now extinct groups must have conferred a decisive competitive advantage, for the earliest angiosperms gave rise to an explosive radiation that produced a majority of today's tropical plant families. By the time the dinosaurs vanished 65 million years ago at the end of the Cretaceous Period, tropical forests not too different from today's were widespread around the globe.

Concurrently with the angiosperm radiation, the world was literally breaking up. For about 100 million years before the Cretaceous, nearly the entire surface of the earth was consolidated in a single contiguous megacontinent geologists call Pangea. There were no major barriers to the dispersal of either plants or animals, and many species enjoyed what were effectively worldwide distributions. The absence of geographical barriers allowed the early angiosperms to spread into portions of Pangea that subsequently became today's separate continents.

This early period of mixing has left its indelible stamp on the world's tropical forests, for plant evolution at the family level has all but stood still during the ensuing 80 million years. Evidence of the Pangean origin of tropical forests can be found in the similar lists of plant families that would be registered in forests in Borneo, Africa, or South America. Alwyn Gentry has demonstrated that the same half-dozen plant families dominate the tree flora of tropical forests the world around. Outside of Southeast Asia, where the dipterocarp family prevails, the legumes (pea family) almost invariably take first place for the number of species representing a single family.

Plant families have maintained their relative rankings with extraordinary consistency over 80 million years in the forests of such (now) disjunct regions as New Guinea, Madagascar, and South America. This degree of consistency can be explained only by some as yet unknown but highly deterministic rule of resource partitioning. The central tenet of the non-equilibrium hypothesis—that success or failure over time is determined only by successive throws of the demographic dice—is glaringly incompatible with the similarity in composition of forests half a world apart.

The conservatism of family representation in tropical forests is affirmed by consistent differences between the montane and lowland tropics, for example. The laurel family (Lauraceae), among others, is most prominent at mid-elevations, so the family composition of a tree plot at 1500 meters in Madagascar more

The mountains of Asia and New Guinea are home to scores of species of rhododendrons, members of the heath family. This one, *Rhododendron fallacinum*, embellishes the slopes of Mt. Kinabalu, Borneo, above 1800 meters.

closely resembles that of a plot at the same elevation in the Andes than it does a lowland forest in the same island. Heaths (Ericaceae) and composites (Compositae, daisy family) appear in abundance at still higher elevations in tropical mountains all over the world, but are nearly absent in lowland floras. Moreover, there are suggestions that certain families are better represented on poor soils, whereas others achieve greater prominence on rich soils, irrespective of continent. That such global parallelisms should persist for millions of years is nothing less than extraordinary.

Stilt rooted trees, such as this *Iriartea deltoidea* palm in Peru, are a common sight in most of the world's humid tropical forests.

Whereas the origins of plant families may date to the time of Pangea, today's species have arisen relatively recently. Particular adaptations, such as buttresses and stilt roots, are possessed by hundreds of species of trees found in all parts of the tropics. As adaptations, these structures improve stability on shallow or water-logged soils, and they are especially prevalent in tall or large-crowned species. Buttresses and stilt roots have evolved repeatedly in numerous evolutionarily independent lineages to reduce the risk of toppling. Therefore, these adaptations generally are examples of convergence. So, too, are the universal similarities in the vertical organization of tropical forests, because the participating species evolved their individual characteristics long after the continents drifted apart. In this case, evolutionary convergence is ensured by the mechanism described in Chapter 5.

The mere fact that geographically remote forests may resemble one another in certain characteristics is not in itself evidence of convergence, for resemblances can arise through parallel descent from common progenitors. The fact that many of the same plant families occur in tropical forests around the world is best explained by common descent from ancestral forms that predated the breakup of Pangea. On the other hand, the similarities of family composition that Gentry has documented are so remarkable as to be mysterious. It is hard to imagine that such consistencies could have been maintained for 100 million years without the action of some powerful constraining force. Similar forces operating in widely separated localities would constitute a form of convergence.

Convergence in the Zoological Realm

Zoologists have been less interested in testing the convergence paradigm, perhaps because some early attempts to communicate the idea through seductive anecdotal imagery were rudely dismissed as artifacts of wishful think-

(*Facing page*) Striking morphological resemblances suggest evolutionary convergence in these rain forest mammals of Africa (*left*) and the Neotropics (*right*). From top to bottom the pairs are: pigmy hippopotamus and capybara; water chevrotain and paca; royal antelope and agouti; yellow-backed duiker and red brocket deer; giant ground pangolin and giant armadillo.

ing. Statistically valid evidence, it was claimed, was lacking. The burden thus fell on those who followed to devise statistically correct procedures for comparing faunas. But this was the least of the challenges on the path to investigating convergence in animals.

A more fundamental dilemma lay in deciding what to measure, because there was no agreement as to what properties of animals might be expected to show evidence of convergence. For animals, it is hard to find examples of a direct selective action of the tropical environment that are as clear as those of the buttresses and stilt roots of plants. Animals are not so closely tied to the physical world as plants because they can often avoid physiological stress by taking appropriate action. For example, animals can moderate their temperatures by basking in the sun, repairing to the shade, or seeking the cool, constant environment of a burrow. Their greater flexibility suggests that the sizes, shapes, and appearances of animals should be selected by factors other than, or in addition to, the physical environment. Since it is not obvious what these other factors might be, the problem seemed inaccessible to most zoologists, and therefore uninteresting.

Moreover, until recently zoologists did not have the benefit of formal sampling techniques, analogous to the tree plots of botanists, to lend statistical credibility to their work. Quantitative data on the structure of tropical animal communities—whether vertebrate or invertebrate—are extremely scarce, even today. For many parts of the tropics, all that are available are species lists, the products of expeditions undertaken to stock the world's museums.

Given the separate evolutionary histories of the continents, it would be naive to expect, for example, that each species of bird or monkey in a forest in Africa would have an identifiable counterpart in South America or Asia. After all, convergence in the botanical realm was at the family level, not the species level. Even convergence at the family level would be impossible for most animal groups, because animals have evolved so much more rapidly than plants. Whereas a majority of the tropical plant families arose in the Cretaceous (130 to 65 million years ago), most families of birds and mammals have been in existence less than half this long. Thus, the tropical continents share relatively few families of vertebrates, despite the striking similarity of their floras. Intercontinental comparisons thus have to be independent of taxonomy; instead, they can be based on functional groupings, such as guilds.

Another statistical problem that plagues the convergence issue is a lack of controls. One looks for similarities, but how similar is similar? It would be surprising if the bird communities of two tropical forests were not more alike, for example, than the bird communities of a tropical and a temperate forest. There is no other frame of reference to serve as a standard. In effect, the lack of controls stands the problem on its head. Since zoologists cannot ask whether two communities or faunas are as alike as predicted by some a priori theory, they are obliged instead to focus on differences. Rather than study convergence, by default zoologists actually study nonconvergence.

One of the first attempts to apply statistical methodology to intercontinental convergence was undertaken by two ornithologists, James Karr and Frances James, who examined forest bird communities in Panama, Liberia, and Illinois, the latter serving as an ad hoc control.

MAMMAL FAMILIES

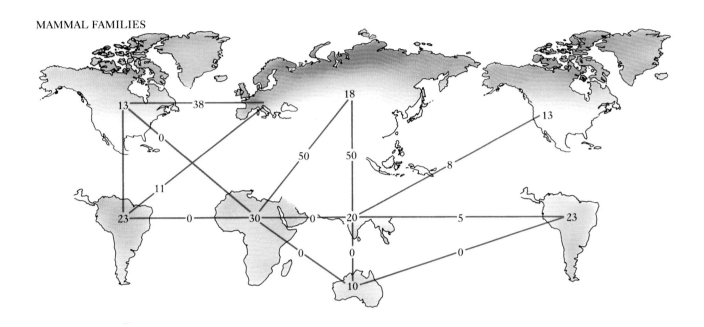

FLOWERING PLANT FAMILIES

The tropical regions share few mammal families but many plant families. Figures within each region indicate the number of families to occur there; figures linking the regions give the percent similarity of the two biotas. "Wandering" mammal families and cosmopolitan plant families have been omitted.

From museum specimens, they obtained measurements of standard morphological features such as bill dimensions, wing length, and tarsus length. Then, using these measurements, they employed multivariate statistical procedures to array the species in a two-dimensional "morphospace." The highly abstract procedure generates a pair of coordinates for each species based on combinations of morphological characteristics. When the coordinates are plotted as points on a two-dimensional graph, species that are similar in morphology will cluster together, whereas the points for dissimilar species will be widely separated. Such plots achieved satisfying groupings of species having similar ecological roles—New and Old World flycatchers, for example—and clear separation of species having distinct ecological roles, even when the species were members of the same family, such as frugivorous and insectivorous hornbills.

In the pattern of clustering of species in "morphospace," the two tropical communities clearly resembled one another more closely than either did the Illinois community. The tropical forests, for example, contained far more birds that walk on the ground. Having relatively short wings and long legs, these clustered together in the morphological plots.

Although many such resemblances were found, there were some conspicuous disparities. Hummingbirds provided the prime example. Hummingbirds are among the tiniest of birds and have exceptionally short tarsi (legs) and long bills. The hummingbirds had no morphological counterparts in the Liberian community, although they had indisputable ecological counterparts in the nectarivorous but nonhovering sunbirds. The comparison thus produced mixed results. Yes, there were distinct resemblances in the morphological organization of the two tropical bird communities, but there were also unexplained differences. Is this enough convergence to satisfy the evolutionary expectation? Who can say?

A different approach to the question was taken by David Pearson, then of Pennsylvania State University. Morphology, he reasoned, might not offer the best expression of convergence. As we noted above, the physical characteristics of animals are not under the same constraints as those of plants because animals can adjust their circumstances at will by choosing appropriate microenvironments. Accidents of history will determine the array of families and genera initially present in any region, and the diversity of morphological forms that can appear during subsequent evolution may be limited by the genetic and morphological features of the ancestral organisms. Behavior, Pearson reasoned, might be more flexible than morphology and more responsive to immediate conditions.

It might therefore be better to examine, in the most direct manner possible, the functions that a group of animals performs in the ecosystem. The functional role of animals is best represented by their feeding behavior. Behavior is the interface between animals and the habitat. Because feeding takes place on, or is directed toward, particular substrates such as bark or leaves, it is more directly related to the structure of the habitat than is a species' morphology. If rain forests around the world are structurally convergent, then the behaviors animals use to harvest their food should reflect the physical similarities of the habitat.

The presumption is a reasonable one because a species' behavior can be far more op-

(Left) An emerald toucanet *(Aulacorhynchus prasinus)* feeds on the fruits of a *Sapium* tree *(Euphorbiaceae)* in a Costa Rican cloud forest. *Sapium* fruits, like many others that are dispersed by birds, open when ripe to reveal bright red arils encasing the seeds. The flashy signal attracts passing birds from afar. *(Right)* Magnified Old World counterparts of toucans, hornbills feed on fruits in the forests of Africa and Asia east to New Guinea and the Soloman Islands. This rhinoceros hornbill of Malaysia *(Buceros rhinoceros)* weighs four times as much as the largest toucan.

portunistic than its morphology. For example, birds with such dissimilar morphologies as hummingbirds, nightjars, woodpeckers, and flycatchers are all capable of catching flying insects on the wing, though not all of them do it equally well. By the same token, a single species, such as the common yellow-rumped warbler of North America, may hop on the ground, picking prey from the surface; glean insects from twigs and foliage; hover at branch tips to pluck berries; or hawk passing insects from the treetops. In other words, species with very different morphologies can at times perform the same functions on the environment, while a single species with fixed morphology can perform many functions. It is this flexibility and independence from morphology that makes behavior a more promising medium of convergence.

Pearson spent two years observing birds in tropical forests in Ecuador, Peru, Bolivia, Borneo, New Guinea, and Gabon. His results provide the best evidence yet obtained for convergence in the zoological realm. Pearson recorded the number of birds he observed per hour using each of nine foraging techniques. Although the bird communities of the six forests differed in their species richness and representation of guilds, they showed striking similarities in their repertoires of foraging techniques. When Pearson ranked the nine foraging techniques in order of decreasing frequency, the rankings obtained in the six forests were highly consistent. In the absence of any normalization of the data, it is remarkable how many of the figures in each category fall within a factor of two. Even without an independent control or an a priori expectation, the similarity in rankings is persuasive evidence for convergence. But the convergence of foraging techniques reflects the structural similarities of the habitat across the sites. It differs from botanical convergence in its indirect relation to the nonbiotic features of the environment.

Nonconvergence in Primate Communities: Learning from the Exceptions

To the small clique of biologists interested in the convergence paradigm, Pearson's result was gratifying, a resounding affirmation of Darwinian theory. But it was not regarded as a breakthrough, because it simply confirmed the expected. Its importance was in the demonstration that behavior converges to a greater degree than morphology.

Given the emphasis on behavior, Pearson's decision to study birds was the right one, because birds are highly diversified in the things they do. But in the overall scheme of things, birds are a minor, almost insignificant component of the community of vertebrate consumers. In the one forest for which such data exist, the total biomass of birds is about 10 percent of that for mammals. Like birds, mammals are diverse: they range in size from shrews to elephants, and they consume almost every possible food resource. As the most prominent group of vertebrates in tropical forests, mammals should provide an important test case for the analysis of convergence in animal communities.

Temperate residents are not accustomed to seeing many mammals when they go for a weekend hike in the woods. A few squirrels perhaps, a chipmunk or two in the right terrain, and a deer if one is lucky. The general impression is that mammals are scarce, and, in relation to most tropical forests, that is true. If one is lucky enough to be able to hike for a day in a tropical forest that has not been hunted to exhaustion, the experience will leave quite a different impression. The hiker will encounter primate troops two or three times an hour. Squirrels of various colors and sizes are also a virtual certainty. Ungulates—deer, pigs, antelope, and their kin—are shy and difficult to observe, but leave conspicuous signs. In addition to these most prominent denizens are many others—sloths, coati-mundis, and agoutis in the Neotropics; porcupines, pangolins, and mongooses in Africa; binturongs, moonrats, and sun bears in Asia. Because it is an island, Madagascar lacks a full complement of mammals, but by day one can see pigs and rodents, and by night various mongooselike predators.

Primates constitute the single most important group of mammals in many tropical forests. Their combined biomass is greater than that of other groups, such as rodents or ungulates. Their dietary habits are extremely varied. Most commonly they eat fruit or foliage, or some combination of the two, but a number of species are more specialized and feed on such items as seeds, bamboo, gum, nectar, or small prey. As primary consumers (feeding on plant parts), primates are more closely tied to plants than are birds, the majority of which are insectivorous and therefore secondary consumers, one ecological step removed from plants.

So difficult is it to observe mammals in the rain forest that little is known about most of them. Primates are the exception. Because many of them are large, diurnal, and live in conspicuous groups, they are relatively easy to detect, even in dense foliage and tall trees. More important, primates are intelligent, observant creatures that can assess the intentions of a person on the ground. Where they are hunted, they become extremely shy, fleeing and hiding at the first hint of an approaching human. But where they are not hunted, and have the opportunity to observe humans going about their daily activities in a nonthreatening way, they become "habituated." Once used to humans, primates will allow observers to approach near enough to record the details of their behavior.

Primates are the passion of primatologists, a curious breed of scientist who is half biologist and half anthropologist. Because their home is in anthropology departments, there are a lot more primatologists than there are specialists in any other group of mammals, despite the fact that there are many more species of bats, rodents, or ungulates. Competition among primatologists for species to call their own has taken them to the farthest corners of the tropics, and it is largely thanks to their efforts that this chapter can be written.

There are four major primate radiations, each of which has filled the arboreal consumer niche of a different portion of the world. One is in the Neotropics, centered in Amazonia; another in equatorial Africa; a third on the island continent of Madagascar; and the fourth in South and Southeast Asia. Each of these radiations has produced 12 to 18 genera and,

This leaf-eating red colobus monkey (*Colobus badius*) is to the forests of Africa what the red howler is to the Neotropical forest. The red leaf monkey (*Presbytis rubicundus*) provides a counterpart in Borneo.

except for Madagascar, about 45 rain forest species.

In comparing the four radiations, we can be confident that any observed similarities are not the lingering vestiges of ancient phylogenetic ties, for not a single genus is shared between any two of the radiations. Of the six possible pairwise comparisons of the four radiations, five are composed of entirely different families. The sixth, that between Africa and

Two sifakas (*Propithecus verreauxi*) relaxing in a typical posture. The thick trunks may feel cool in the day's heat. Sifakas, along with the lemurs, are representative of the remarkable radiation of prosimian primates in Madagascar. Once widespread, their mainland relatives disappeared at the end of the Eocene, some 40 million years ago.

Asia, includes members of two families, the Pongidae (apes) and Cercopithecidae (macaques and guenons). The Asian and African branches of these two families probably separated in the Miocene Epoch, some 15 to 20 million years ago. Otherwise, the lineages in the four regions have been separated since the Oligocene Epoch or before (35 million years ago), the time when the very first monkeys appear in the fossil record. We can thus be assured that the four assemblages represent independent throws of the evolutionary dice, and that their respective adaptations evolved as responses to local or regional environmental conditions.

Thanks to the intrepid efforts of a generation of primatologists, primates have been widely censused in forests all over the world, so there exists an ample store of information on their population and community structure. Even a casual look at the data reveals that the primate communities of different tropical regions fail to show the expected convergence. Instead, there are differences in a number of basic features. For example, the biomass of primates in different forests varies over an order of magnitude. The highest naturally occurring values are probably in central and west Africa, but data are scanty because there are almost no effectively protected forests in this region. More extensive data are available for the Neotropics and Southeast Asia. The average biomass in Asia is about 50 percent higher than in the Neotropics, although even the best sites attain barely more than a third of the value recorded for the Kibale Forest in Uganda—2525 kilograms per square kilometer. Unfortunately, there are no data for any rain forest site in Madagascar, although a dry forest at Morondava in the western part of the island supports the remarkable total of 2720 kilograms per square kilometer, one of the highest ever recorded. It is clear, therefore, that primate communities do not converge in biomass; the differences are attributable primarily to widely varying numbers of large-bodied folivores.

With the exception of Asia, the numbers of species of primates coexisting in the forests of the four regions are remarkably similar. Madagascar, perhaps as a consequence of its insular status and relatively small area, contains fewer rain forest primates than any of the other regions. Nevertheless, diversity at the community level is on a par with the best mainland sites; there are up to 12 species in some areas. The diversity of Madagascar's primate communities was originally even higher, but about a third of its species went extinct shortly after man first arrived about 1000 years ago. Local diversities in central Africa and Amazonia are similar, and the richest localities contain about 14 species. Asia clearly lags behind with no more than 7 species present at prime sites in West Malaysia and Sumatra. The deficiency, as we shall soon see, is entirely attributable to a paucity of small forms.

Of the characteristics we shall examine for evidence of convergence, species diversity is the one most likely to have been molded primarily by the evolutionary history of the several regions. We shall neglect species diversity for the moment, however, in order to follow Pearson's lead in studying attributes that better reflect the interactions of the animals with their immediate environments. To that end, we shall next consider diets and feeding behavior.

A Digression on Primate Diets

Primates are particularly flexible in their feeding habits, and they cannot be classified into guilds so easily as birds. Instead, primatologists recognize a series of rough divisions that can be used to relate each species to a dietary spectrum extending from specialized folivory at one extreme to specialized insectivory at the other. At the two extremes are radically different digestive adaptations. Folivores eat large quantities of low-quality food material, which they process slowly in long, elaborate digestive tracts. In contrast, insectivores consume high-energy, high-protein food in small packets that are processed quickly in short simple guts.

Folivores and insectivores are further distinguished by important differences that arise as consequences of metabolic scaling. For the technically minded, the metabolic rates of all mammals from mice to elephants, plotted against body mass, conform closely to a double log relationship of slope 0.75. This is to say that small animals require proportionately (though not absolutely) more food than large ones. This rule, known as Kleiber's law, has important implications for ecology, for it imposes size constraints on animals having different diets. Folivores, for example, are usually

(Left) The large sacculate stomach of this African *Colobus* monkey, much like the rumen of a cow, aids in fermenting a leafy diet. Food passes slowly through the long convoluted intestine, as microbes assist in the release and transformation of nutrients. *(Below)* In contrast, the angwantibo *(Arctocebus calabarensis),* a nocturnal resident of the same African forests, is largely insectivorous and possesses a much simpler digestive system that passes food relatively quickly.

large relative to species that consume food of higher nutritional value. Large animals have slow metabolisms, and a slow metabolism is needed to extract the scant nutrient content of fibrous leaf material. Protein and other important constituents are released gradually through the activity of enzymes, frequently aided by symbiotic microorganisms. Food is processed at a leisurely rate in the long and capacious guts of folivores, which serve to maximize the time over which enzymes and microbes can assist in digestion.

Large body size confers two further advantages on a folivore. First, by Kleiber's law, less food must be consumed for each kilogram of body weight. Other things being equal, food can pass more slowly through the gut. Second, a slower passage time enables the digestion of poorer-quality food material. Thus, by being large, a folivore effectively increases the proportion of edible foliage in its environment. Small folivores, such as voles and lemmings, are able to eat only tender young shoots of the highest protein content, while elephants can subsist on almost any kind of vegetable matter, including stems, twigs, and bark.

Insectivores possess a diametrically contrasting set of digestive adaptations. Their food is of the highest quality, but in comparison to green leaves it is extremely scarce. Intake is limited by the rate at which prey can be found and captured, not by digestion, and in most cases large body size does not result in higher rates of prey capture. This being so, it is advantageous for insectivores to be as small as possible in order to minimize their daily food requirement. Folivores and insectivores thus fall at the poles of an adaptive spectrum based on metabolic and digestive properties.

Frugivores and omnivores lie along the middle portion of this spectrum. Frugivores, like folivores, often have special digestive adaptations that allow them to consume immature fruits protected by what are called plant secondary compounds. These compounds are toxic substances that often inhibit the digestive processes of plant-eating animals, both vertebrates and invertebrates. Immature fruits often contain such substances, which include alkaloids, terpines, and glycosides, as well as the more universal tannins. Secondary compounds are the source of plant-derived pharmaceuticals such as reserpine, strychnine, quinine, and curare. When fruits ripen, these deterrent molecules are converted to harmless or tasteless derivatives. That is why green apples can produce indigestion, whereas ripe ones do not.

Specialized plant eaters are often capable of detoxifying secondary compounds in their livers, and in consequence they gain access to food supplies that are poisonous to other species. Although such ability can confer clear competitive advantages, the detoxification capacities of most vertebrate plant eaters are limited, as is our own. We are able to detoxify ethyl alcohol, for example, but only at a limited rate. A folivorous primate may similarly be able to consume small quantities of a number of toxic substances, but it may be unable to handle large quantities of any of them. Thus vegetarian primates typically eat the fruits or foliage of several to many different plant species every day.

Frugivores that are able to detoxify the defensive compounds of immature fruits are frequently able to consume foliage as well, because plants use similar substances to protect both their fruits and their leaves. This flexibility of diet blurs the distinction between folivores and frugivores, since, as we shall see, primates belonging to both categories readily vary the proportions of fruit and foliage in their diets in accordance with the relative availabilities of these resources in the environment. Thus, the distinction between the two dietary categories is somewhat arbitrary, and folivores are defined as species whose diets contain more than 50 percent leaf material, while, over the course of the year, frugivores consume more fruit than leaves.

The case of omnivores is different. Omnivores are defined as species that include a significant amount of animal prey in a diet com-

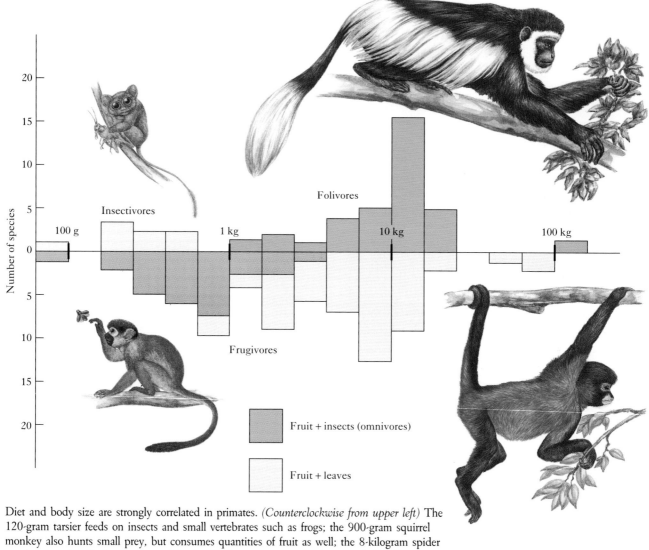

Diet and body size are strongly correlated in primates. *(Counterclockwise from upper left)* The 120-gram tarsier feeds on insects and small vertebrates such as frogs; the 900-gram squirrel monkey also hunts small prey, but consumes quantities of fruit as well; the 8-kilogram spider monkey is primarily frugivorous; while the 12-kilogram guereza is almost entirely folivorous.

posed largely of plant material. The latter supplies most of the caloric content, and prey provide most of the protein requirement. The manner in which the protein requirement is satisfied defines a critical adaptive watershed. Since protein contains about 14 percent nitrogen, and nitrogen is a scarce element in many environments, especially in the tropics, plants tend to protect their most proteinaceous parts with secondary compounds. An animal that is to obtain its protein from plant sources must therefore be adapted to detoxify secondary substances. To do so, an animal must possess the suite of characters described above—relatively large body size, slow metabolism, long gut, and slow passage time.

These traits are antithetical to the ability to satisfy a protein requirement with small prey, which is best achieved through small body size. With few exceptions, small body size carries the cost of a relatively rapid metabolism and its corollary, a fast passage time. Omnivores thus tend to be considerably smaller than vegetarian species, and to be limited in the types of plant materials they can digest. The typical fare consists of sugary, often mildly acidic fruits that are high in water and low in fiber—the type of fruit we humans prefer to eat. Such fruits are digested easily and can be passed quickly from the gut, leaving space for a slow but steady intake of insects and other small prey. Since their calories derive mostly from fruit, omnivores can be larger than specialized insectivores, but they seldom weigh more than 3 to 5 kilograms as adults.

Dietary Habits in Four Primate Radiations

The ecological properties of geographically isolated primate communities can be expected to converge if tropical forests around the world produce similar arrays of fruit, foliage, and animal prey, the principal food resources of primates everywhere. To test this prediction, we shall examine the distribution of dietary specializations in representative forest-dwelling primate communities of the Neotropics, Africa, Madagascar, and Southeast Asia.

In all four communities primates are the predominant group of arboreal consumers. The comparisons of the communities could be intrin-

sically biased if that role were usurped by other organisms in one or more of the regions. Clearly it is not. Squirrels are prominent in African and Asian forests, as are civets, and marsupials (opossums) and procyonids (raccoon relatives) are common in the Neotropics, but the biomass of none of these groups, nor all of them together, is equal to more than a minor fraction of the primate biomass in undisturbed forests. We can thus be assured that the comparisons are reasonably fair.

The dietary habits of primates in the four regions differ in many respects. Leaf eaters are poorly represented in the Neotropics but make large contributions to the biomass of primate communities elsewhere, particularly in Asia. On that continent, every species, except the small, nocturnal lorises and tarsiers, consumes some amount of foliage. African communities show good balance across the spectrum of dietary specialization, while omnivores are overrepresented in Neotropical communities. The rain forest primates of Madagascar resemble those of Asia in that they are more folivorous and less frugivorous and omnivorous than those of the other regions.

Since body size and diet are adaptively related in primates, the distribution of body sizes in the four assemblages reflects the dietary pattern. The strongly folivorous African and Asian primates are generally of large size. For example, of the nine diurnal species at Makokou in Gabon, only one, the talapoin monkey, weighs less than 2 kg as an adult, and at Ketambe in Sumatra none of the seven species is this small. In contrast, small species predominate in the Neotropics. At Cocha Cashu in Peru, five of nine diurnal species weigh less than 2 kg. All but one feed heavily on insects, more so, appar-

Artist's conception of the extinct *Megaladapis,* a gorilla-sized primate that occupied Madagascar until it vanished shortly after humans arrived on the island.

Nocturnal primates are found in all four regions, and with no exceptions they are small (less than 2 kg). The nocturnal habit is most prevalent in Madagascar, where nearly half the species are active at night, and least prevalent in the Neotropics, where there is only one, the night monkey, *Aotus.* Perhaps these differences are the result of historical accident, because nocturnal primates fill roles in the impoverished mammal fauna of Madagascar that in Amazonia are filled by numerous species of marsupials (opossums) and procyonids (kinkajou, olingo). Africa and Asia harbor intermediate numbers of nocturnal forms, most of which are partly to mainly insectivorous. Once again, the New World proves anomalous, because it has spawned an extensive radiation of small diurnal omnivorous species for which every Old World counterpart is nocturnal. These daylight-feeding omnivores are the tamarins and marmosets, none of which weighs more than 1 kg. Why such creatures should flourish in Neotropical forests and not elsewhere remains a puzzle. Like the orangutan, they should not exist, but do.

Why are there more folivores in Asia and Madagascar, a more evenly balanced mix of species in Africa, and so few folivores and so many small omnivores in the Neotropics? The finding of so many differences between the communities contradicts our hypothesis of similarity due to convergence, but suggests no alternative principle. The obvious implication is that the nature of the food supply available to primates varies strongly between regions, but how and why? In order to find some possible answers, we shall examine more closely the food supply at one well-studied locality.

ently, than any Old World species. Again, the New World primates prove to be different. As for Madagascar, it must be kept in mind that the recently extinct forms were all large, one of them the size of a gorilla. The four to five extant diurnal species all weigh more than 2 kg, in keeping with their strong tendencies to feed on leaves.

How Some New World Monkeys Cope with Scarcity

A few years ago, with the aid of several collaborators, I conducted a year-long study of primate ecology at Cocha Cashu in Peru's Manu National Park. Out of a community of ten species, we selected five for intensive study, all of them omnivores. These were two capuchins, the brown and white-fronted; the squirrel monkey; and two tamarins, the saddle-backed and emperor. To document feeding and ranging behavior, our study group continuously observed a troop of each species for 20-day periods during each of the four quarters of the year. Concurrently, we measured fruit fall from the canopy by weighing the fruit that dropped into 150 wastebasketlike "fruit traps."

In the seasonal climate at Cocha Cashu, with seven wet and five dry months, the forest experiences a pronounced annual cycle. Although the change of seasons is not as visually spectacular as at midlatitude, many trees lose their leaves during the dry season and then, as in our spring, leaf out, flower, and fruit when the rains return. The dry season thus provides a strong physiological signal that serves to synchronize plant reproductive activity. Consequently, the availability of fruit fluctuates markedly through the year, peaking in the early to mid wet season and reaching a prolonged minimum at about the time the rains end. Thirty times as much fruit may be available at the peak of the fruiting cycle as in the trough. It is therefore interesting to know how consumers such as primates react to the variable food supply.

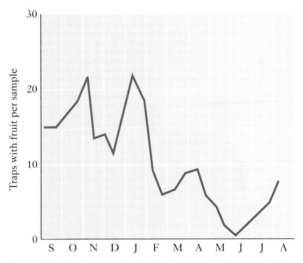

Fruit fall in the forest at Cocha Cashu is markedly seasonal. For 7 to 9 months a year fruit is in surfeit, but for the remaining months frugivorous animals are obliged to depend on other kinds of food resources, often ones of low nutritional value.

Our year-long investigation began in August toward the end of the dry season. For the next nine months, all five primate species ate fruit pulp and almost totally ignored other plant parts, even though seeds and young foliage were usually plentiful. They also searched for small prey while traveling between fruit trees, a practice followed throughout the year. The simple routine of alternating foraging sessions with visits to fruit trees held through April, but gave way to unexpected new behaviors in May, June, and July as the dry season intensified.

Our fruit traps collected almost nothing in these months, and the behavior of the monkeys reflected the evident scarcity. For example, during 21 days in January when fruit was abundant, the group of 12 brown capuchins fed in more than 225 fruiting trees for a total of 2350

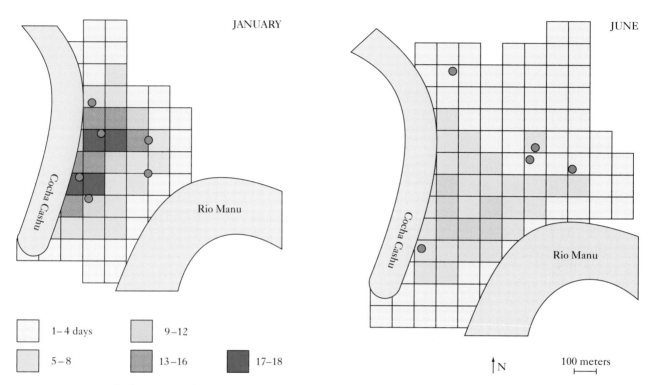

JANUARY

JUNE

Cocha Cashu

Rio Manu

Cocha Cashu

Rio Manu

| | 1–4 days | | 9–12 |
| | 5–8 | | 13–16 | | 17–18 |

↑N

100 meters

The brown capuchin monkey *(Cebus apella)* at Cocha Cashu ranges more widely in June (dry season) than in January (rainy season). In January the troop fed in more than 225 trees while remaining near the center of its home range, whereas in June it searched an area twice as large to find only 50 trees with fruit. The solid circles indicate the locations of trees that accounted for 50 percent of the total time the troop spent feeding on fruit.

minutes. In June, although the troop searched an area twice as large (73 hectares instead of 34), it found only 50 producing trees adequate to support 890 minutes of feeding. By this measure, the animals' fruit intake had dropped to only 38 percent of the January level. Even more extreme reductions were noted in the tamarins.

Observations made during the dry season demonstrated that each species had a previously unsuspected ace-in-the-hole for surviving the annual period of fruit scarcity. The capuchins

switched to palm nuts, the tamarins to nectar, and the squirrel monkeys to insects. In each case, the ability to subsist on these alternative food resources depended on a particular physical attribute of the species. The strong jaws of the capuchins give them a jowly, bulldog appearance, but also give them the 140 kg of bite force needed to crack the nuts of *Astrocaryum* palms. None of the other primates in the community is able to crack these nuts.

When the fruit supply failed in their territories, the tamarins camped out for days at a

(Left) Tightly packed, like the kernels on an ear of corn, the nuts of a Scheelea palm are difficult to extract from the unbroken cluster. The dominant male is the only member of a capuchin troop strong enough to break off the first one. *(Right)* In addition to feeding on palm nuts, this brown capuchin *(Cebus apella)*, like the tamarin, depends heavily during the dry season on nectar supplied by the flowers of this canopy liane *(Combretum assimile)*.

time in the tangled crowns of trees bearing a robust liana, *Combretum assimile.* Conserving energy by moving only when necessary, they would fan out every few hours to lap nectar from showy spikes of orange-red flowers. There were weeks in July when up to 90 percent of their food intake came from *Combretum* flowers. With a sugar content of only 8 to 10 percent, *Combretum* nectar offers pitifully little substance to a 450-gram animal—rather like trying to subsist on a few glasses of orange juice. In fact, our data show that the tamarins lose 10 to 15 percent of their normal weight during the *Combretum* season, yet if not for *Combretum*, it is hard to see how they would survive. They are able to do so only by virtue of being small, smaller even than our common gray squirrel.

Rather than living within defined territories as the tamarins do, squirrel monkeys range over areas that are extremely large for a forest monkey, up to 10 square kilometers. At first their wanderlust seemed anomalous and inexplicable, but later it became apparent how ranging far and wide helped these monkeys to survive the period of food scarcity.

An emperor tamarin *(Saguinus imperator)* pauses to scan for predators in its search for frogs and katydids in a Peruvian forest. Red stains on its Fu Manchu moustaches from the pollen of *Combretum assimile* provide telltale evidence of the role of tamarins as pollinators of this plant.

Among the very few kinds of trees to ripen fruit crops during the lean months from May through July are strangler figs. Stranglers are a curious tropical phenomenon. They are parasitic plants that germinate in a crotch or knothole high in the crown of a host tree. By starting out at the top of the canopy, stranglers gain immediate access to sunlight. From this position of advantage, they drop long, thin aerial roots to the ground. Once these roots penetrate the soil and begin taking up nutrients and water, the strangler grows very rapidly, develop-

ing a crown that soon overtops that of its host. At this stage, thicker roots begin to envelop the host's trunk, hence the name strangler. After smothering its host, a strangler can become a free-standing tree. Some of these stranglers develop prodigious dimensions—heights may be over 50 meters and spreads in excess of 40 meters.

When one of these giants fruits in the dry season, birds and monkeys congregate from all directions. Usually flocks of *Brotogeris* parakeets discover the bonanza, and in their noisy zeal draw the attention of primates from hundreds of meters away. I have seen over 100 monkeys feeding simultaneously in the crown of such a tree.

Giant stranglers play an important role in the lives of squirrel monkeys and other forest animals because they may provide the only significant source of fruit for many weeks. Giant stranglers, however, are rare in the forest, and fruiting giant stranglers are rarer still. At any given time there may be only one, or none, in several square kilometers. It is to find these rare trees that we think squirrel monkeys roam so widely. When one comes into fruit, several troops will appear within a day or two and feed in alternation.

During the dry months it is either feast or famine, for if there are no ripe fig crops in the vicinity, the squirrel monkeys may be obliged to go for a week or ten days without eating fruit. Being too large (800 to 1000 grams) and too numerous to survive on nectar, and too small to crack palm nuts, they cannot make use of the food supplies of capuchins and tamarins. Instead, they ceaselessly hunt insects all day long. Just as tamarins are too large to be hum-

During the dry season, squirrel monkeys *(Saimiri sciureus)* may range over as many as 1000 hectares
(4 square miles) in search of figs. Hundreds of species of strangler figs occur in all parts
of the tropics. Brightly colored fig fruits such as those on the right are typically dispersed
by birds and diurnal mammals; dull green and brown figs are normally dispersed by bats.

mingbirds, squirrel monkeys are too large to be insectivores. The largest birds to live exclusively on a diet of insects weigh no more than 300 grams. Being larger than this confers no advantage in increased capture rates; the large size only entails an extra metabolic cost. Although we have not captured and weighed squirrel monkeys as we have tamarins, it is reasonable to assume that they are unable to maintain their body weights when living entirely on small prey. Even though they catch about one insect per minute, most of their prey are tiny and provide very small rewards. To survive through intermittent periods of fruit deprivation by resorting to full time insectivory, squirrel monkeys would be at a severe disadvantage if they were any larger than they are. Again, making it through the lean season is accomplished by being small.

This realization provided the answer to why there are so many small primates in the Neotropics, but failed to answer other questions, such as why folivory does not seem to be an option for New World primates, and why there are few or no small omnivorous primates in the Old World. To fill in these missing pieces of the puzzle, we had to take a broader perspective.

The Role of Keystone Plant Resources

Once we were alerted to the idea that dry season feeding behavior might hold the key to understanding the adaptations of rain forest animals, additional facts began to fall into place. Generations of students and professional biologists had studied a large and ever growing number of vertebrate species at Cocha Cashu, including not only primates but also peccaries, procyonids, some terrestrial rodents, some arboreal marsupials, and a variety of frugivorous birds. What we might never have known by studying primates alone gradually became obvious as we accumulated knowledge about these other species. The fallback food resources that were first identified in the primate study—palm nuts, nectar, and strangler figs—were in fact heavily used during the dry season by nearly every vertebrate frugivore in the forest. Peccaries eat palm nuts; so do squirrels and agoutis. Night monkeys, kinkajous, opossums, and many birds compete with tamarins for *Combretum* nectar. Howler monkeys, spider monkeys, titi monkeys, as well as capuchins and more than a score of bird species gravitate to fruiting stranglers. Palm nuts, nectar, and figs therefore play a key role in the forest ecosystem, and have accordingly been termed "keystone plant resources."

Despite many years of observations, we have found very few additional plants that produce edible fruits, seeds, or nectar in the season of scarcity. Our list of these resources now stands at only a dozen species. While 12 species might sound generous in the context of a temperate forest, it does not in the context of a tropical forest containing more than 1000 plant species. About one percent of the plant diversity thus supports up to 80 percent of the animal biomass, and appears to establish the ecological carrying capacity of the environment. These few species of plants are extremely important not only to our understanding of tropical forests but also to our management of them, a matter to which we shall return in the final chapter.

Even in the humid tropics survival in a seasonal environment requires passing through an annual bottleneck. In the north, most species lay eggs and die, migrate, go into diapause, or hibernate. Only a comparative few manage to remain active. In the tropics, production is not shut down for so long, nor are climatic conditions so severe. Most vertebrates remain active, although in Madagascar there are primates that go into torpor for the length of the dry season. Those remaining active must often alter their diets. In the absence of the preferred food supply, they must make do with nutritionally inferior products that provide a bare minimum for survival, like the nectar used by tamarins. But even such poor-quality alternatives are actively sought by competing species.

We found, for example, that the tamarins harvest no more than 2 percent of the supply of *Combretum* nectar, and even that much only under rather special circumstances. The remaining 98 percent is taken by an array of other monkeys and a score of bird species. The birds are especially formidable competitors because their pointed bills enable them to draw the nectar in the flowers down to the level that tamarins, with their relatively clumsy tongues, are unable to remove any more. The persistence of tamarins in the ecosystem is thus dependent on the slimmest of margins. Their survival, as stressed above, is made possible by their small size and commensurately modest food requirements. If not suitably equipped by morphology or behavior, a species cannot succeed in the face of intense competition for a limited supply of resources. Success can be achieved only through the possession of adaptations that provide access to one or more of the keystone resources.

Keystone Resources in the World at Large

If a mere dozen plant species play such a crucial role in the ecosystem at Cocha Cashu, it is reasonable to wonder whether there are not equally important species in other tropical forests. However, the exceedingly sparse literature on this point refers to less than a handful of localities. Few investigators have looked simultaneously at both the whole plant community and the whole animal community and have examined how the two communities interact, per-

haps because the breadth of study and the coordination required are daunting. Fortunately, the information we do have comes from three of the four main regions in our comparison.

Barro Colorado Island is the only New World site other than Cocha Cashu that provides a comparison. Although there has been no directed effort to identify keystone plant resources on BCI, there have been year-long studies of several of the principal species of mammals, and from these studies it is possible to extract a somewhat hazy overview. As at Cocha Cashu, the primates seem to depend heavily on figs during the dry season, while squirrels, terrestrial rodents, and peccaries feed on palm nuts. Nectar does not seem to be important. Otherwise, the pattern is similar to the one at Cocha Cashu.

For what is known about keystone resources in the forests of Southeast Asia, we are indebted to Mark Leighton, a gifted naturalist who has spent many years in Borneo. Leighton studied hornbills and their feeding behavior in the Kutai Reserve. Even though the climate was less seasonal than that in southeastern Peru, the production of ripe fruit varied markedly from season to season and year to year. During low points in the cycles of abundance, a few plant species attracted large numbers of avian and mammalian frugivores, including hornbills, primates, and squirrels. Prominent among these plants were strangler figs. The rest were other species producing fleshy fruit. Palms were infrequent in the forest at Kutai and did not play any special role, nor did Leighton observe the seasonal use of nectar by species that normally consume other resources. When little

Starting from a tiny seed in the crown of its host, a strangler fig lowers roots until it envelops its host in a deadly embrace. Although considered pest trees by foresters, stranglers play important roles in tropical forest ecosystems as "keystone plant resources" for many birds and mammals.

fruit was available, some hornbills left the area entirely, presumably migrating to other parts of the island, while primates and squirrels resorted to leaves or immature fruits. Thus the ability to survive periods of food scarcity in Borneo does not depend on small size or extraordinary bite force, but rather on an ability to migrate or to cope with plant secondary compounds.

Our knowledge of keystone resources in Africa is similarly limited to a single site, the French research station at Makokou in Gabon. A major program extending over two decades has generated nearly all the knowledge in existence about the central African forest. Fruit production at Makokou is cyclical, peaking in the wet season and declining to an annual low in the major dry season, as at Cocha Cashu. Figs, however, are rare. They make up only a few percent of the diets of primates, in contrast to the heavy use they receive at Cocha Cashu (up to 70 percent of dry season diets). Instead, three trees appear to play the role of keystone resources—*Polyalthia suaveolens*, *Coelocaryon preussi*, and *Pycnanthus angolensis*. These trees fruit regularly in the dry season at Makokou, producing fruit in moderate to high abundance. Annie and Charles Gautier, G. Dubost, A. Brosset, and their associates have documented the consumption of these species by at least four monkeys, six terrestrial rodents, four squirrels, all seven ungulates, three turacos (large, brightly colored frugivorous birds), and four hornbills. Again, as in Borneo, there is no indication that palms or nectar play special roles in this forest. Instead, many of the primates increase their use of foliage during periods when fruit is relatively scarce.

The Timing of Fruiting and Leafing Cycles

Tropical forests around the world experience marked rhythms of plant activity. Because one can find flowers or fruit at any time, the impression is created that food is always available. Yet careful measurements reveal that fruit production varies strongly with the seasons, even in the most monotonous climates on earth. But, as we saw in Chapter 1, not all tropical climates are the same.

Climates having one rainy and one dry season in a year are widespread in the Neotropics, both north and south of the equator. In central Africa, the humid equatorial zone generally supports a two-peaked climate, with two rainy and two dry seasons in a year. Finally, the Malaysian region is wetter than the other two and includes large areas that experience no regular dry season.

It seemed plausible that these three climatic regimes could result in major differences between the regions in the timing of fruiting, flowering, and leafing cycles. This is a possibility I recently explored with my Duke University colleague Carel van Schaik, whose tropical experience is in Sumatra. The hypotheses we tested related to whether leafing and fruiting cycles were synchronous or nonsynchronous.

Unlike their Old World counterparts, few New World primates regularly consume leaves during the dry season, whereas every species at Makokou and at several Asian localities consumed moderate to high amounts of leaves, at least during one season. These facts seemed to imply that folivory was not an adaptive option open to New World primates. That could be the case if the seasonal cycles of fruit and leaf production ran in phase in Neotropical forests. If the two cycles were in synchrony, the two resources would be scarce at the same time, a situation that would not encourage switching from one to the other. If, on the other hand, the two cycles ran consistently out of phase, so that fresh foliage was abundant when fruit was scarce and vice versa, switching would be a predictable evolutionary response. The Gautiers had already demonstrated switching in central African primates, providing a hint that primates will alternate between fruit and leaves when the cycles are out of phase. Finally, fruiting cycles in the Asian forest are notoriously irregular, and there is great variability from year to year in the amount produced. An irregular burst of fruiting every few years would show no correlation in a statistical test with the regular seasonal pattern of leafing. Under these circumstances, most animals would not be able to subsist on fruit alone, but fruit could be consumed by folivores whenever it was available.

Fortunately, the data we needed to test these predictions were to be found in the literature, thanks mainly to the efforts of numerous primatologists who had documented the food supply available to their study animals. Simultaneous monthly measurements of fruiting and leafing activity over one or more yearly cycles were available for four localities, one in the Neotropics (BCI), one in central Africa (Makokou), and two in Southeast Asia (Kuala Lompot in West Malaysia and Ketambe in Sumatra). Straightforward correlation analysis

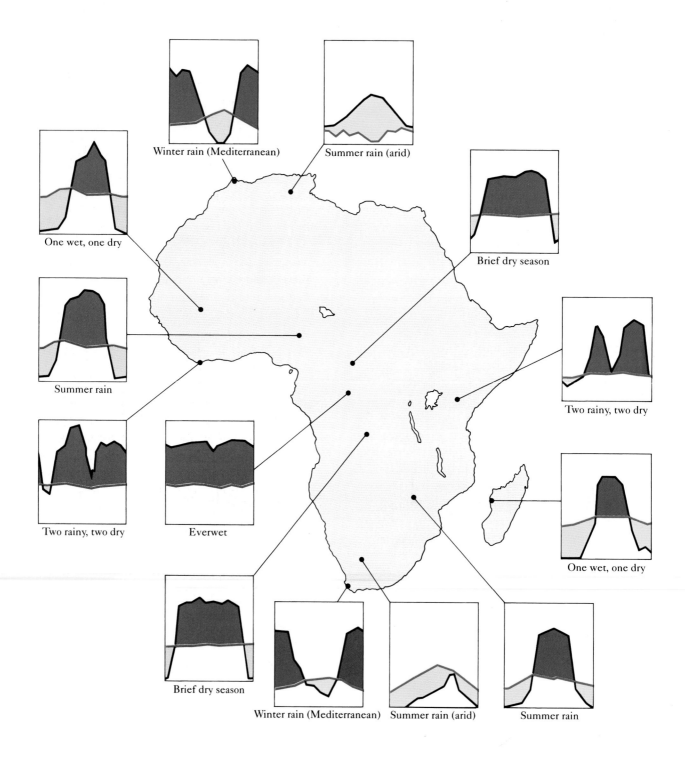

Winter rain (Mediterranean)

Summer rain (arid)

One wet, one dry

Brief dry season

Summer rain

Two rainy, two dry

Two rainy, two dry

Everwet

One wet, one dry

Brief dry season

Winter rain (Mediterranean)

Summer rain (arid)

Summer rain

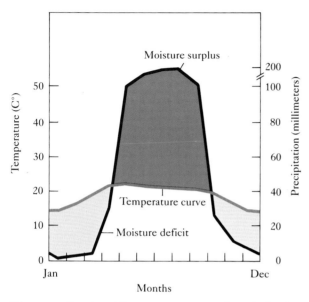

The seasonality of rainfall increases symmetrically on both sides of the equator in Africa, giving rise to a progression of climates: everwet; single brief dry season; two wet and two dry seasons; one long wet and one long dry season; arid climate with summer rain; "Mediterranean" climate with winter rain.

showed that the fruiting and leafing cycles at these localities were, as anticipated, respectively in phase, out of phase, and uncorrelated.

Because the leafing and flowering cycles are synchronized at BCI, a frugivore there does not have the option of switching to fresh foliage when fruit is scarce. Instead, as we noted earlier, the primates at BCI and other Neotropical localities must rely on figs, palm nuts, nectar, and small prey to tide them over the season of scarcity. Given that a large animal would starve on a diet of nectar or insects, it becomes evident why so many New World primates are small. The small body sizes of many species, and the scant evolution of folivory, account for the low biomasses of New World primate communities.

The opposite situation occurs in the central African forest at Makokou. When fruit is scarce, young foliage is plentiful, and vice versa. Furthermore, there is no time in the year when one food type or the other is not fairly abundant. Taken together, these results account for the pronounced seasonal switching between fruit and foliage observed in the primates at Makokou, and the exceptionally high biomasses of some African primate communities.

As we had anticipated, fruiting and leafing activity were uncorrelated at the two Asian localities. Intervals between fruiting episodes are long and the supply of edible foliage is irregular and often low. We can assume therefore that periods of scarcity are frequent and prolonged. At such times, Asian primates must increase their intake of mature leaves or immature fruit, feats that require relatively large body size. Here we have a plausible rationale for the reduced diversity of Asian primate communities, and the absence of small diurnal primates. A sporadic availability of fruit can further explain the absence of specialized primate frugivores, such as the mangabeys of Africa and the spider and wooly monkeys of the Neotropics. The absence of omnivorous species in addition to specialized frugivores helps to explain the low biomasses of Asian primate communities.

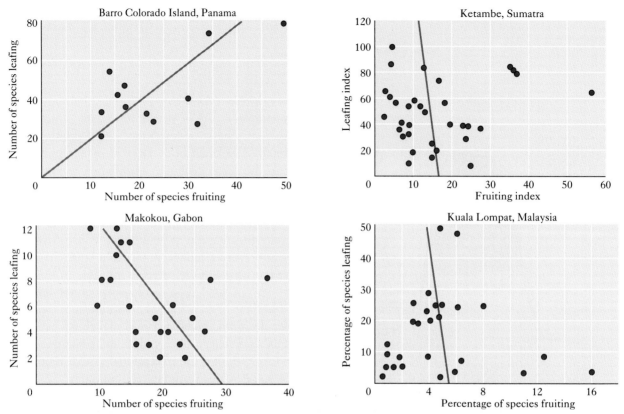

In a tropical forest in Panama, young leaves and fruit are scarcest at the same time of year; in Gabon, fresh leaves are most available when fruit is scarce, and vice versa. In the two Southeast Asian localities, there is no significant correlation between leafing and fruiting because leafing occurs throughout the year and fruiting is concentrated in irregular bursts every few years.

A Reaffirmation

Inspired by the success of botanists in demonstrating many points of structural convergence in tropical forests around the world, and imbued with faith in the Darwinian paradigm, zoologists began seeking evidence of convergence in tropical animal communities. Hopes were at first raised by Pearson's discovery of a striking concordance in the foraging behavior of tropical forest birds on four continents. It seems likely that the similarities in behavior can be attributed to the very structural similarity of the forests to which the botanists had directed our attention. Avian foraging behavior relates directly to the structure of the habitat because it is the habitat that provides the substrates upon which and within which birds search for prey.

Duly encouraged, we chose as the next step to investigate primates as examples of primary consumers. The road at this juncture suddenly became tortuous, for in place of convergences we found the contrary. The primates of different continents are not at all similar: they differ in size, diet, and abundance. Our null hypothesis—that all tropical forests would produce similar arrays of primate food resources—

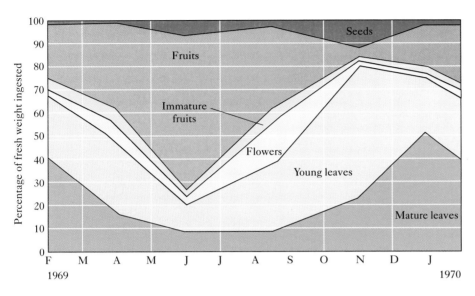

The gray langur consumes fruits, young leaves, and mature leaves in very different proportions as the seasons progress in a dry forest at Polonnaruwa, Sri Lanka. The langur prefers fruits when they are available.

proved to have been much too simplistically constructed. We found that subtle differences in rainfall patterns, perhaps scarcely noticeable to the everyday resident, can exert major influences on the timing of leafing and fruiting cycles.

In retrospect it was extraordinarily naive to have expected evolution to conform to vague assumptions about benign and equable climates or superficial resemblances in the structure of the habitat. Both of these matter, but in the proper context. To primary consumers, these features of the environment are not the most crucial to finding enough food to grow and reproduce.

The timing of resource cycles proves to have ultimate significance, for it determines whether the environment is perceived as relatively uniform or as gyrating through boom-and-bust cycles. A given environment can support only the array of adaptations that can compete successfully for keystone plant resources during seasons of scarcity. Subtle differences in the seasonal availability of fruit and foliage that can be detected only by statistical analysis emerge as a key factor in the evolution of primate community structure.

We have now come full circle. Nonconvergence has its Darwinian explanation, as does convergence. The physical environment acts indirectly through plants to shape animal characteristics and capabilities in ways that are becoming predictable, but the results reported here represent only a beginning. We must now wonder, for example, whether the community organization of animal groups other than primates has responded in similar ways to variable patterns of resource abundance. Scientists should not hesitate to attack such questions, for the time remaining to compare intact animal communities in farflung tropical forests may not be long.

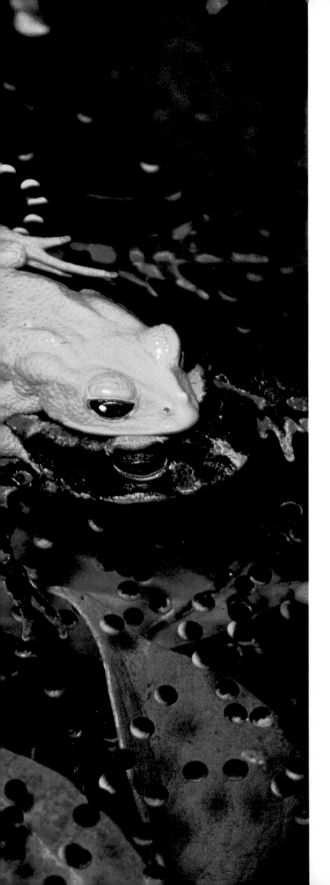

8

Conserving Biodiversity

Disillusioned by the bloodshed of World War II, a group of American Quakers decided to emigrate once peace returned. In keeping with their abhorrence of war, they elected to go to the one country in the hemisphere that, by popular mandate, had abolished its military establishment. In Costa Rica they recognized a country of peace and civility.

These latter-day pioneers established a community they named Monteverde (Green Forest) on a remote mountaintop in the central portion of the country, where they cleared the virgin forest to establish modest dairy farms and, later, a cheese factory. The combination of a near-perfect climate, easy access to undisturbed cloud forest, and a hospitable, English-speaking community of

Two pairs of golden toads (*Bufo periglenes*) in amplexus as the females (*dark*) deposit their eggs in a mountain stream.

nonhunters began to attract more Americans, but this time they were biologists.

Among the many discoveries made by a growing corps of researchers was the golden toad. Often featured in Costa Rican travel posters, the golden toad, along with the resplendent quetzal, is one of the biological jewels of Monteverde. Unlike the drab and warty toads familiar to most Americans, the orange-red color of the golden toad is so intense that the toads appear to glow with an incandescent force in the misty shadows of the cloud forest. So far as is known, Monteverde is the only place in the world where the golden toad is found. A decade ago the species abounded at Monteverde, living along the ridgetop near the many little rivulets that coalesce into roaring torrents on the steep slopes below.

Two or three years ago worried biologists began to complain that they could no longer find the golden toad in its usual haunts. More recently, intensive searches have failed to uncover any at all, and it is feared that the species may now be extinct. The mysterious disappearance of the golden toad is one of a number of instances around the world in which biologists have documented the decline or total loss of amphibian populations. In many cases the declines cannot be attributed to any identifiable cause.

The case of the golden toad exemplifies the growing number of challenges to the nascent discipline of conservation biology. Practitioners of this burgeoning interdisciplinary field are attempting to apply the tools of ecology, genetics, and population biology to the problems of preserving genetic diversity and preventing extinctions. Conservation biologists seek to deduce the causes of extinction in the hope that a better understanding of this process will lead to improved programs for protecting endangered species and habitats.

The number of extinctions the world is experiencing is unknown even to within an order of magnitude. Some conservation organizations will claim hundreds, even thousands of species lost every year, but such assertions are, at best, wild guesses. In truth, the status of less than 1 percent of all species, most of them large mammals and birds, is accurately known. Many of the rest have yet to be scientifically described. Nevertheless, there is abundant cause for serious concern about the future survival of biodiversity on this planet, for every year nature loses ground faster than the year before, especially in the tropics.

Deforestation in the Humid Tropics

Tropical forests, including both evergreen and seasonal formations, once covered nearly 25 million square kilometers of the earth's surface. Humans have been gradually encroaching on this environment since our species began to emerge from the hunter-gatherer stage thousands of years ago. Hundreds of millions of people now occupy the humid tropics, concentrating in regions where fertile soils permit sustainable agricultural development. Among the most densely settled portions of the tropics are the alluvial plains of the Ganges in India, the Mekong in Southeast Asia, and the Niger in Africa; the volcanic islands of Java, Sumatra, and the Philippines; and the volcanic highlands of Central America, Cameroon, and East Africa. As human populations have expanded in these and other regions of the tropics, the area covered by evergreen forest has fallen to less than 8 million square kilometers. Perhaps as much as half of the world's tropical forests had already been lost by 1990. This land is now occupied by cropland, plantations, pasture, or shifting agriculture, or it has simply been exhausted and abandoned as wasteland.

If tropical nature is going to survive, an adequate amount of land will have to be protected within the 8 million or so square kilometers of evergreen forest that still remains. We are at a timely juncture at which to project the future survival of tropical forests, for in 1990 two independently conducted studies were released that evaluated rates of deforestation worldwide. A report issued by the Friends of the Earth (FOE) estimates current deforestation at 142,000 square kilometers annually, an area the size of Florida or Greece; one issued by the World Resources Institute (WRI), in collaboration with the United Nations, puts the annual loss at 160,000 to 200,000 square kilometers, roughly equivalent to the state of Washington or the country of Syria. Both studies are based on a country-by-country interpretation of satellite imagery. The estimates are not identical because the evaluations use somewhat different criteria for what constitutes deforestation; for example, they differ on whether to count as deforested areas that have been massively logged or converted to second growth.

Both surveys document major increases in the rate of deforestation since 1980, when similar reports were released by the Food and Agriculture Organization of the United Nations and the U.S. National Research Council. FOE found that the annual rate of deforestation had doubled between 1980 and 1990, from 0.9 percent to 1.8 percent of the forest currently

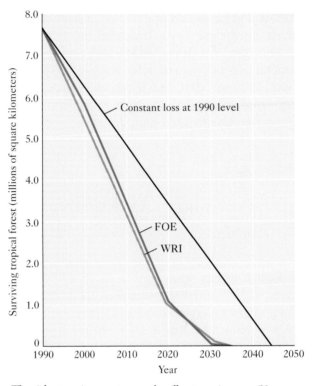

The richest environment on earth will not survive even 50 years unless drastic measures are instituted to curtail deforestation. The straight line projects the global decline in humid tropical forest if we assume no change after 1990 in the amount of forest cleared annually; the other two lines represent projections based on data from Friends of the Earth and the World Resources Institute.

remaining. WRI found that the 1980 rate had increased 50 percent by 1987, leading to a doubling of the rate in 12 years. The somewhat different estimates of the two organizations for the annual loss in 1990, and for the rate of increase in the annual loss since 1980, are in opposite directions and tend to cancel. The two estimates consequently yield very similar projections; both predict that 90 percent of the remaining forest will be lost in 33 years. This loss would be in addition to the amount already lost before 1990. If current trends were to continue, the annual loss would peak at around 250,000 square kilometers in the period from 2005 to 2010, and would then begin to fall simply because not enough forest would remain to sustain such high declines.

The United States: An Example not to be Emulated

Although it may seem hard to imagine that regions as vast and unpopulated as the central Amazon and Congo basins could ever be completely deforested, one has only to recall the liquidation of hundreds of millions of acres of virgin forest in the United States. That country lost most of its forests in a few decades under less pressure of population than exists in the tropics today.

For more than 150 years the U.S. government vigorously promoted expansion of the western frontier and settling of the continent. The Homestead Act of 1862 was a landmark piece of legislation designed to accomplish this national goal by distributing free land to all comers. Recognizing that the growth of commerce depends on an efficient transportation infrastructure, the government made land grants totalling 183 million acres to stimulate railroad construction in the West. Like much legislation in tropical countries now decried by conservationists, the Homestead Act confers title to the land only after the installation of "improvements."

In their zeal to reap the bounty of a virgin continent, Americans succeeded in 100 years, between 1870 and 1970, in clearcutting 500 million acres of virgin forests east of the Great Plains; plowing more than 99 percent of the tallgrass prairie; draining over 90 percent of the prairie wetlands in Illinois, Iowa, Minnesota, and several other central states; converting much of the shortgrass prairie to sagebrush or mesquite scrub by systematic overgrazing; and damming, diverting, polluting, or channelizing nearly every major stream and river in the coterminous 48 states. Operating under few political or legal restraints, and often with the active participation of the government, Americans transformed the natural history of an entire continent.

Species that once predominated in the great biomes have been reduced to pitiful remnants or have disappeared altogether. The passenger pigeon, alleged to have been the most abundant bird on earth at the time of Columbus, was driven to extinction by unrestrained overexploitation. It was later followed into oblivion by the Carolina parakeet and ivorybilled woodpecker, victims of the wholesale logging of southeastern forests. The common white-tailed deer was so thoroughly extirpated in the eastern states by unregulated hunting that the species did not exist in the Shenandoah National Park when the park was created in the 1930s. To redress this embarrassing deficiency, the Park Service was compelled to restock with animals transported from another state. Plant life has suffered as well. The most valuable tree of the East, the American chestnut, was devastated by a fungal blight inadvertently introduced from Europe. Weeds of alien origin now dominate the view along roadsides and in fields and pastures across the land.

Seventeenth-century European settlers in North America led a way of life strikingly similar to that of the slash-and-burn agriculturists who are responsible for most of the deforestation in the tropics today.

Because there are no jobs in the cities, this Peruvian family is forced to eke out a meager existence in the Amazonian forest, more than a day's travel from the nearest doctor or secondary school.

In the plains states, vast numbers of grouse, shorebirds, and waterfowl succumbed to land conversion, overhunting, and the draining of wetlands. Several species of shorebirds are now so rare that mere sightings are prized by birdwatchers. The American bison, once the most abundant large mammal on the continent, was reduced in about 40 years from an estimated 60 million to the brink of extinction. The elk, bears, and wolves that John James Audubon saw in abundance on his excursion up the Missouri River in the 1830s are now but dim memories.

When one considers the grand scale on which nature has been abused in North America, it is remarkable that extinctions have not been far more numerous. Since Europeans began to settle the continent, the United States has lost approximately 1 percent of its birds, mammals, reptiles, amphibians, and vascular plants. Nevertheless, the list of threatened and endangered species is long and growing. That more extinctions have not occurred already is attributable to the low species diversity of temperate plant and animal communities, to the vast extent of the principal biomes of the continent, and to a long history of disturbance by fire and glaciation.

In contrast, extinctions have been rampant in Hawaii, the only part of the United States to lie within the tropics. More than 50 bird species that occupied the archipelago before the arrival of humans have disappeared forever, and half of those that remain are now so rare that their continued survival is in doubt. In the now completely altered lowlands of the Hawaiian Islands, the only birds to be seen in most places are alien species introduced from other parts of the world.

Which of these two models—continental North America or Hawaii—best represents the tropics of the future? This is a completely open question that will be decided by action or inaction within the life spans of most people now alive.

The Vanishing Tropical Rain Forest

The deforestation now occurring over most of the humid tropics is not very different from the deforestation that occurred in North America over the past century and a half. Human populations are expanding rapidly, and governments are vigorously promoting the exploitation of natural resources and the colonization of frontier zones. Virgin forests are being wastefully timbered, or they are being converted into pasture and cropland that last only a few years before the soil wears out. Hitherto inaccessible wildlands are being actively explored for petroleum and mineral deposits. And governments are pushing roads into the last remaining wilderness to encourage the settling of what are viewed as lands going to waste. Propelling all this activity is an intense desire to achieve a decent standard of living and to "catch up" to the personal prosperity of the developed world.

Given that the peoples of most tropical countries are far behind those of the developed countries in education, raising living standards through innovation and technology is out of the question. Instead, the only viable option is to exploit natural resources, a measure made all the more urgent by onerous debt burdens and by limited opportunities for competing in the global market. Hard currency markets for cof-

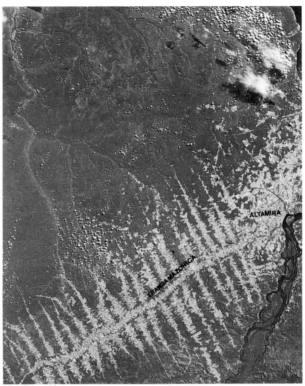

(Left) Shortly after construction was completed, the Transamazonian highway near Altimira, Brazil, stretches through virgin forest. *(Right)* A satellite view of the same area a few years later reveals an advancing wave of deforestation. Lots are parceled out to farmers along perpendicular access roads.

fee, sugar, chocolate, and other tropical products are growing at less than 1 percent per year, while the producing countries must expand their outputs by 2 to 3 percent per year just to stay even with population growth. Thus, many countries have only one means of boosting foreign exchange earnings: to expand exports of raw materials, including timber, oil, and minerals. At best, this tactic will be pursued only during a period of transition to more sustainable forms of economic activity. At worst, what is happening now will make it impossible to attain sustainability later.

We are not far from the year 2005, the year of no return for tropical forests if the projections of the FOE and WRI reports prove accurate. Reversing the trends documented in

these reports will require nothing less than a revolution in the way natural resources are managed by both governments and individuals. However, if one searches human history for examples of revolutions in attitude, 15 years does not seem enough time. One has only to consider the 72 years it took to gain women's suffrage, from its proposal at Seneca Falls, New York, in 1848 to the passing of the Nineteenth Amendment; the 100 years from the end of the Civil War to the passing of the Civil Rights Act; or the 70 years the Soviet Union foundered under communism before the party leadership acknowledged that the system did not work as intended. These and other examples of the reluctance of people to alter their views and ways of life do not bode well for the tropical forest. If biodiversity is to survive for posterity, the world will have to experience a change in attitude more rapid than ever experienced before.

Reducing the pace of tropical deforestation, and eventually halting it altogether, will be doubly difficult because the immediate forces driving the process vary markedly from one country to another. A variety of different approaches and solutions will therefore be required, although the ultimate force driving deforestation—the pressure of expanding human populations—is a factor everywhere. A few examples will illustrate the point.

Brazil

Possessed of more tropical forest than any other country, in 1988 Brazil led the world in land deforested with a loss of 50,000 square kilometers. The amount is roughly two-thirds of the total global loss of 75,000 square kilometers in 1980. A variety of pressures are threatening forests in the Brazilian Amazon.

The construction of a paved highway into the western state of Rondonia unleashed a land rush of historic proportions in the 1980s. In less than 10 years, more than a million small farmers immigrated into the region. A satellite

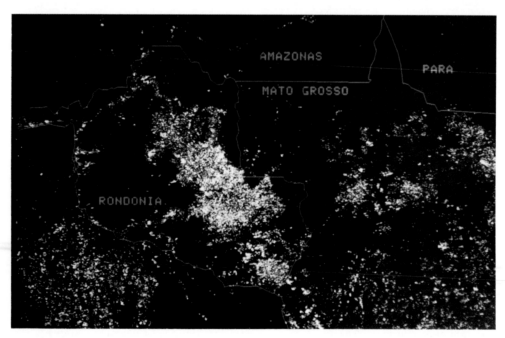

This composite satellite view of Rondonia at night shows the fires of slash-and-burn cultivators during six weeks at the end of the 1987 dry season. More than 6000 fires were counted on a single night. The thin lines of fires stretching across northern Rondonia and southern Amazonas reveal sections of the Transamazonian highway system. The scene is approximately 1400 kilometers across.

Mining for cassiterite, a tin ore, in Brazil. Forest destruction, stream contamination, and the displacement of indigenous peoples by mining is becoming a severe problem for the Brazilian government, even in the sparsely populated Amazon.

image taken at night in the dry season of 1988 showed over 6000 fires burning simultaneously, as landowners fired tracts in anticipation of the annual rains. Every year Brazil loses an estimated 2 billion dollars worth of timber to burning.

Thousands of unruly gold miners have illegally invaded the Yanomamo Amerindian reserve in the northern territory of Roraima and are poisoning scores of streams with mercury used to extract the precious metal. The government of President Fernando Collor de Mello has called on the air force to bomb the miners' landing strips in a so far unsuccessful effort to dislodge them.

Huge areas of Matto Grosso, Pará, Maranhão, and other states have been deforested in government-sponsored schemes to promote cattle ranching. To this end, Brazilian taxpayers have provided $2.5 billion in subsidies to 460 politically connected large operators. However, the ranchers have profited more from the subsidies and land speculation than from cattle raising.

Giant hydroelectric projects such as this one at Tucuruí on the Rio Tocantins in Brazil have flooded hundreds of thousands of hectares of virgin tropical forest and displaced native peoples.

Lacking petroleum, Brazil is now manufacturing gasohol from sugar cane. To supply a growing market, Brazil has vastly expanded the area sown to sugar cane in Piauí, Ceará, and other states. Again, because of a lack of alternative domestic energy sources, Brazil is constructing hydroelectric dams on several Amazonian tributaries, and is planning a series of additional projects that will inundate millions of hectares of tropical forest in Pará, Amazonas, and neighboring states.

As the huge Carajas project gains momentum, one can look ahead to the wholesale deforestation of Pará, a state nearly twice the size of Texas. Carajas is the site of the world's largest deposit of high-grade iron ore—18 billion tons of it—enough to sustain mining at current rates for 250 years. Vast amounts of charcoal

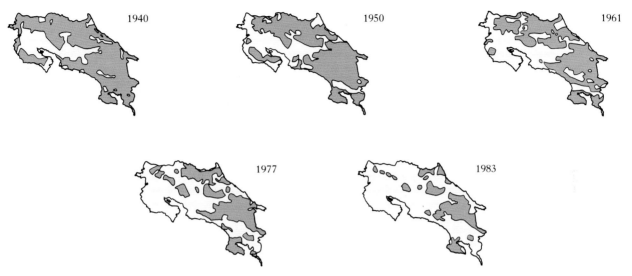

The progressive deforestation of Costa Rica. Having reserved nearly one quarter of its national territory, Costa Rica has done more than any other tropical country to conserve its natural resources. Ironically, however, its deforestation rate is one of the highest in the Americas, and little forest is predicted to remain outside of protected areas by the end of the decade.

are required for smelting the ore. A recent study by the Grande Carajas Interministerial Commission projected a demand for charcoal that would consume 1000 square kilometers of forest annually. The Carajas project, like the construction of the road into Rondonia, received substantial financial support from the World Bank, notwithstanding legal requirements for environmental impact statements.

As if all these pressures on the Brazilian Amazon were not enough, timbering on a large scale has hardly begun. Currently, world demand for tropical hardwood is being met with production from Southeast Asia, particularly Malaysia and Indonesia. However, both these sources will be largely depleted before the de-

cade ends. It is anticipated that Amazonia will then become the principal source for a market that is expected to grow steadily into the next century.

Ivory Coast and Nigeria

As far back as 1975, 14 countries of sub-Saharan Africa held populations too large for their cultivatable land to support, given the subsistence methods in use. These countries account for one-third the land area and one-half the population of sub-Saharan Africa. Since 1975, the Sahelian drought of the mid-1980s has turned hardship into catastrophe. The Ivory Coast is just one of several West African countries in which forests are bearing the brunt of the Sahelian drought. Within a period of five years, a tier of countries just to the north—Mauritania, Mali, Niger, Burkina Faso, Chad—lost thousands of square kilometers of once productive land to desertification. Millions of people were displaced from their ancestral homelands. Hundreds of thousands of them have been resettled into camps as permanent wards of the United Nations. Millions of others have migrated south into better-watered neighboring countries.

The Ivory Coast alone received an estimated 1.5 million Sahelian refugees. For that country, the influx represented the equivalent of an overnight invasion of the United States by 30 million destitute aliens, a population greater than that of its 20 largest cities combined! Once the Sahelians had moved into the Ivory Coast, they had few options. Most quickly headed for the only unoccupied land available—the country's forests. There they slashed and burned to create subsistence plots. The FOE report estimates that 16 percent of the remaining forest cover is vanishing every year. If deforestation continues at this rate, the Ivory Coast's forest estate will be completely eliminated by 1995.

A similarly early demise of Nigeria's forests is in prospect. With more than 115 million inhabitants, Nigeria is losing 14 percent of its remaining forest cover each year, as slash-and-burn cultivators push back the last frontiers. By its own admission, the government is powerless to restrain the tide, and it anticipates that all remaining forest will be eliminated by the mid-1990s.

Central Africa

The rain forest of central Africa shelters one of the most diverse and interesting mammal communities on earth. Gorillas, chimpanzees, elephants, duikers, and the okapi, a strange short-necked giraffe, are a few denizens of this forest. Fortunately, poor transportation has so far prevented wholesale logging in this region, although the recent completion of the Trans-Gabon railroad undoubtedly presages future trends. Human populations are still relatively low as a result of poor soil and high death rates in the past from disease. Any hope for an African rain forest in the future lies here.

The immediate threat is logging, illustrated by the case of Zaire. That country contains the world's third largest expanse of tropical forest after Brazil and Indonesia. Because the costs of transportation are high, only the most valuable species are exported as timber. In their efforts to harvest these few species, loggers follow particularly wasteful practices: up to 25 trees are destroyed for each log exported. Regardless, the government is planning to increase timber exports from 150,000 cubic meters in 1984 to 5,000,000 cubic meters by 2000.

Madagascar

The world's fourth largest island is an evolutionary relictuary. Since Madagascar broke away from the East African mainland about 100 million years ago, a widening water barrier has impeded the invasion of more recently evolved lineages. Madagascar's isolation has left

This once forested landscape lies in ruins, unfit for man or beast. Environmental disaster had overtaken much of Madagascar long before the advent of chain saws and bulldozers. Similar scenes are appearing in dozens of other countries, as deforestation exposes fragile soil to erosion.

With a population that is growing more than 3 percent every year, Madagascar has lost most of its forest through slash-and-burn agriculture and systematic burning in the dry season. Despite the government's efforts to prevent further losses, deforestation continues unchecked because the people have no alternative. The ancient soils have been weathering for more than 100 million years and are among the worst on earth. Once they have been cleared, burned, and farmed for a single meager crop of upland rice, they are completely exhausted. What grows back is a pitiful low scrub that persists for years. Unlike other parts of the tropics, where the forest rebounds vigorously after each slash-and-burn cycle, in Madagascar clearing the forest is essentially an irreversible step. The loss of woody vegetation has led to widespread erosion of the inherently infertile soil and some of the most severe environmental degradation on earth. The remaining patches of evergreen forest, nearly all perched on steep slopes, are being reduced by 8 percent each year. At the present rate of attrition, nothing will be left by the turn of the millennium, and the world will have lost an unrepeatable evolutionary experiment.

Thailand

As recently as 1950, 70 percent of Thailand was covered by forest. However, a 1988 satellite survey revealed to a startled government that forest cover had decreased during the interim to 15 percent. In the same year, a particularly heavy monsoon washed down deforested slopes, carrying floods into the lowlands, de-

it with a flora of nearly 10,000 endemic species and a fauna more typical of ages past. Madagascar's famous lemurs, for example, are living relatives of the adapids, among the earliest of primates, creatures that abounded some 50 million years ago in the Eocene of Wyoming.

stroying crops, and leaving 40,000 homeless. In response to the disaster, the government decreed a ban on logging, notwithstanding the expected damage to a major sector of the economy. The government did not anticipate, however, that its action would have the perverse effect of tripling the price of lumber in Bangkok. Spurred by this incentive, timbering has resumed illegally on a larger scale than ever before. Aerial surveys conducted by the Forestry Department in early 1989 reported 54 percent more forest logged than in 1988, before the ban went into effect. Landless peasants often take advantage of logging roads to enter otherwise inaccessible regions, and so it was no surprise when the survey also documented a 28 percent increase in the amount of forest cleared for slash-and-burn agriculture. Unless the government is able to intervene more effectively than it has to date, all remaining forest in Thailand will disappear in nine years.

Just as drought in the Sahel increased the pressure on the forests of neighboring countries, the logging ban in Thailand has induced a sharp increase in timbering in adjacent Myanmar. Much of the cutting is taking place in remote areas occupied by ethnic minorities who do not recognize the central government.

Indonesia

Possessing the second largest area of tropical forest after Brazil, Indonesia is rushing to maximize current revenues by expanding timber exports. The principal beneficiary of this one-time bonanza is a pervasive military establishment that effectively controls the nation's forests. Deforestation is estimated at 12,000 square kilometers per year. About half this loss is attributable to slash-and-burn agriculture, 30 percent to government-sponsored development projects, and the remainder to destructive logging.

These brief sketches cover a geographically dispersed sample of countries, and in the aggregate reveal the depth of the deforestation problem. Although the immediate causes may differ from country to country, in each case powerful forces are promoting deforestation: overt government development policies, often backed up by subsidies and tax breaks; legal requirements that land must be cleared to be titled; powerful business interests that regard forests and other natural resources as sources of quick profits; a widespread reluctance to recognize the land claims of tribal peoples; the desire of many governments to settle remote frontier zones out of fear of encroachment by neighboring countries; a lack of alternative energy sources that compels people to cut forests for fuel; an insatiable global market for tropical hardwood; and ultimately the simple but overwhelming force of ever increasing human numbers, as manifested in the hundreds of millions of would-be farmers who have nowhere to go for land but ever deeper into the forest.

It is not easy to see how the pressure generated by all these forces can be deflated. As an observer of human affairs, I can only regard the situation with extreme pessimism; as a scientist, I can rise to the challenge of conservation biology, which is to discover how nature can do with less and less, and still somehow survive. It is in the guise of optimist that I shall now continue with a speculation on how humanity and nature might continue to coexist into the next century.

Will Parks Preserve Nature?

Within a few decades, unperturbed nature will cease to exist outside of protected parks and reserves. Parks offer the best prospects for preserving the flora and fauna of the tropics. The question we must answer is, Will parks succeed in preventing extinctions? With less and less land remaining each year in a condition meriting protection, it is clear that the rate of creation of new parkland will decline in the future. Although the number of protected areas worldwide doubled over the last two decades, the total area under protection grew more slowly.

As of 1990, most countries had designated less than 3 percent of their national territories as nature reserves, a figure that falls far below the 10 percent target urged by the International Union for the Conservation of Nature. Barely a dozen countries have protected more than 10 percent. Two that have are Luxembourg, with nearly 40 percent, and Botswana, with 20 percent, neither a bastion of tropical forests.

In countries that do contain substantial areas of tropical forest, nature reserves are typically in mountainous districts or in other areas offering limited potential for development. An example is the Grand Savannah of Venezuela, a plateau of breathtaking scenery underpinned by some of the most barren substrate on earth. The United States has pursued a similar policy, protecting 22 percent of Alaska and practically nothing in the fertile breadbasket of the central plains.

The understandable reluctance of politicians to withhold valuable natural resources from productive use is likely to result in undesirable, if unintended, consequences. One can point to West Malaysia, which on paper might

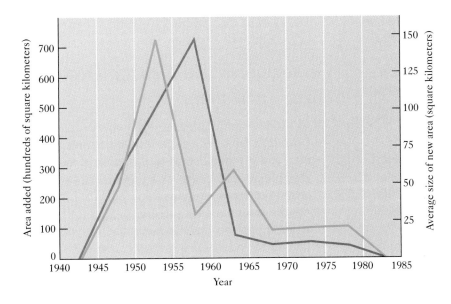

The annual total area (*green*) and the average size (*red*) of parks and reserves established in East Africa since 1940. As wildlands continue to shrink, the worldwide trend is for fewer and smaller new parks.

A relative of the common wild boar, the bearded pig (*Sus barbata*) of Southeast Asia pursues a distinctive nomadic lifestyle in search of mast (seeds), a habit that brings the herds into frequent contact with humans. Exposure to hunters plus fragmentation of its forest habitat make this species a top candidate for early extinction.

seem to have an exemplary park system. But closer inspection reveals that nearly every hectare of protected land is in mountainous terrain. Politicians, beset by conflicting demands from their constituents, often attempt to provide something for everyone. In siting parks in the mountains, Malaysia's government overlooked the fact that the Asian megafauna, with its elephants, tapirs, sambar deer, rhinoceroses, bearded pigs, sun bears, gibbons, and wild oxen, is quintessentially a feature of lowlands. These animals simply have not been given enough space of the right kind to survive in West Malaysia, and some local extinctions seem inevitable.

We now arrive at the main issue, which is, How much space will be necessary to maintain such charismatic large vertebrates into the next century and beyond? Conservation biology has taken several quite distinct approaches in the attempt to answer this crucial question.

Demographic and Genetic Death of Small Populations

Loss of habitat, as well as other circumstances, may cause a population to decline in number. In general, as the size of a population decreases, the risk of extinction rises. Extinction can be forestalled by preserving areas of habitat large enough to include minimum viable populations of endangered species. To specify how large an area should be protected, one needs to know the species' natural population density and the size of the minimum viable population.

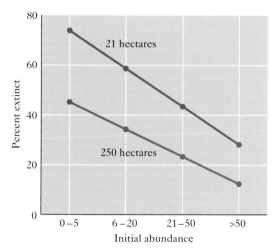

Uncommon bird species were much more likely than common ones to disappear from two forest fragments in the state of São Paulo, Brazil. More species survived in the larger fragment at all levels of abundance. In this and other studies, rarity has proven the best predictor of vulnerability to extinction.

The problem of determining the minimum viable size of any particular population is complicated by normally occurring fluctuations. The populations of all organisms fluctuate from year to year, or generation to generation, some characteristically more than others. The less stable a population, the more likely it is to suffer demographic extinction when, by chance, the number of individuals falls so low that recovery does not occur. The minimum population size that will ensure long-term viability is thus a matter of the inherent tendency to fluctuate. It will consequently be much greater for some species than for others. Since the extent to which particular populations fluctuate in nature is generally not known, this approach does not offer much practical help.

A related approach is to estimate the minimum number of breeding individuals required to maintain a population's genetic diversity. The calculations rest on assumptions about the species' breeding system, sex ratio, year-to-year variance in population size, and mutation rate. Although the theory behind such calculations has been developed to a very sophisticated degree, the fact remains that quantitative information is needed to satisfy the equations, and that information is generally not available. Most practical applications of the theory have been directed to the design of breeding programs for endangered species in zoological parks. As for unmanaged wild populations, geneticists have opined that 300 reproductive individuals should be sufficient to maintain genetic diversity in species with more or less conventional breeding systems.

While this magic number provides a useful rule of thumb for managers, and a scientific arguing point in Congressional hearings on endangered species, a difficulty arises in using it to determine the size of any particular conservation area. How can we decide on a certain number of square kilometers when the spatial requirements of different species vary so widely? An area quite sufficient to support a population of 300 individuals of an herb or an insect might support only one tree of a rare species, and it might be simply inadequate for a bear or a bobcat. This qualification applies to an even greater degree in the tropical forest, where so many species are inherently rare.

The animals that coinhabit the Amazonian forest, for example, differ by more than six orders of magnitude in the amount of space needed to support a single reproductive unit. (A reproductive unit can range from the tens of thousands of individuals in an ant or termite colony, to a troop of monkeys, to a pair of

These giant otters (*Pteroneura brasiliensis*) are to the aquatic realm of the Amazon as the jaguar is to the terrestrial realm. A single group of these highly social otters, containing five to seven individuals, consumes more than 30,000 fish a year.

songbirds.) At the bottom end of the spatial scale are insects, frogs, rodents, and other lowly creatures that attract little international attention. At the upper end are species likely to be pictured in travel brochures. Among these are top predators, such as jaguars, pumas, and harpy eagles, that are compelled by the laws of energetics to be much less common than their prey, and nomadic frugivores, such as fruit crows, macaws, and white-lipped peccaries, which travel great distances between distinct forest types that offer seasonally abundant food. If we were to apply the rule of thumb on minimum viable population size to one of these glamour species—the jaguar, let us say—we would find that a population of 300 breeding adults needs no less than 7500 square kilometers. By this criterion there are only a few parks on earth that contain enough space for jaguars.

If all species lived independently, such that the presence or absence of any one had no effect on the rest, it might not matter that some required more space than others. Those with exorbitant needs could be the expendable ones in a world no longer able to afford the luxury of maintaining jaguars, tigers, and elephants. The rest of nature could proceed about its customary business, mindless of the missing minority. As we shall see later in the chapter, however, scientific evidence is mounting that species are not independent of one another. Despite their relative rarity, the principal predators and frugivores in particular play key roles in maintaining the biodiversity of tropical forests.

Islands as Laboratories of Conservation Biology

Theoretical calculations offer one approach to forecasting the survival of populations confined to parks of limited area. Islands offer a more empirical approach. Islands provide analogs of the parks and nature reserves of the future, which will come to be similarly isolated as surrounding lands are preempted for human purposes.

Geologically old islands, such as those that punctuate the world's seas, are beyond the normal reach of dispersing plants and animals. Their faunas and floras contain only those species (or their evolved descendents) that one way or another succeeded in crossing the water barrier. Island plant and animal communities therefore contain relatively few species, which together are only a sampling of the organisms present on the mainland.

Biogeographical theory holds that in old oceanic islands the number of species is at equilibrium. There is a dynamic balance between the rate at which new species succeed in colonizing and the rate at which established populations go extinct. Large islands typically support more species of a given taxonomic group (such as birds or lizards) than do small islands, because colonizations are more likely to be successful, and extinctions less frequent, on

a landmass capable of supporting larger populations. For islands of a given size, those that are more isolated will be colonized by new species at a slower rate. The slow rate of colonization indirectly lowers the extinction rate by reducing the frequency with which new colonists outcompete, prey upon, or otherwise eliminate established residents. Far islands thus tend to support small numbers of relatively old populations, whereas near islands support larger numbers of more recently arrived populations.

Islands at equilibrium provide a static view of the effect of area on diversity. Simple empirical plots called species-area curves display the numbers of species of birds or other groups known to occur on islands of different size. Such plots are typically log-linear over several orders of magnitude of island size. Whether one studies birds in the Pacific or reptiles in the West Indies, the slopes of such plots fall within a relatively narrow range. The slope of a typical plot indicates that at equilibrium, a 90-percent reduction in area results in a 50-percent reduction in diversity.

We may now begin to deduce the implications of allocating only 3 to 5 percent of the tropical world to nature preserves, especially when it is unlikely that all the protected land in a given country or region will be in one piece. A more realistic assessment is that the largest parks will not exceed 1 percent of the area of the Amazon or Congo forests. During the period required to reach equilibrium, we can anticipate that three-quarters of the species of these biomes will become extinct.

One naturally asks, How long does it take to reach equilibrium? Fortunately, the evidence indicates that it takes a very long time. Esti-

An artist's reconstruction of a few of the many extinct birds that occupied the Hawaiian archipelago for millenia before the Polynesians arrived about 1000 years ago. The bird on the left is a member of the ibis family, and the two on the right are flightless rails. Nearly half of Hawaii's birdlife disappeared before Europeans ever saw the islands, and another quarter has become extinct since then.

mates of these so-called relaxation times are derived from studies of "land-bridge" islands, bits of continental shelf that became islands about 12,000 years ago when melting continental glaciers fed a rapid rise in sea level of 120 meters. For much of the preceding 100,000 years, these landmasses had been integrated into the adjacent continents, and during that long interval their floras and faunas had presumably reached an equilibrium. Among the world's land-bridge islands are Great Britain, Ceylon, Taiwan, Trinidad, and Newfoundland, to name a few of the most familiar.

To estimate the number of species (of birds, bats, lizards, etc.) present on a land-bridge island at the time of isolation one can use the number of species currently found in an equivalent-sized area of the adjacent mainland. The number of presumed extinctions to have occurred since isolation is the difference between this estimate and the number of species now inhabiting the island. For small islands of less than 1000 square kilometers, more species have disappeared than now remain; for larger islands, current species counts tend to exceed the number of presumed extinctions.

Estimated Loss of Species after Isolation of Five Neotropical Land-bridge Islands

Island	Number of species lost in 10,000 years	Extinction coefficient	Number of species lost		Percent lost in first century
			First 100 yrs	First 1000 yrs	
Trinidad	144	1.6×10^{-7}	2	22	0.6
Margarita	246	1.0×10^{-6}	10	80	3.2
Coiba	172	8.8×10^{-7}	5	45	2.2
Tobago	218	8.9×10^{-7}	8	63	2.6
Rey	179	1.7×10^{-6}	8	63	3.7

We can attain a better understanding of the kinetics of species loss by exploring the relationship between the number of presumed extinctions and island area. Biologists have constructed mathematical models based on a variety of assumptions about the rates of species loss through time; these models can generate estimates of an "extinction coefficient" for any particular island. The higher the coefficient, the greater the rate of extinction. The greatest predictive power has been achieved with models specifying that the rate of species loss is proportional to the square of the number of species currently present. This assumption acknowledges that competition between species may hasten extinction. Mathematically, the array of possible two-way competitive interactions in a community can be crudely represented by the number of pairs of species present, which, in turn, is roughly proportional to the square of the number of species. This model predicts that extinctions will occur rapidly following isolation of a landmass, and then ever more slowly as the system asymptotically approaches equilibrium. The extinction coefficients calculated for each island are inversely correlated with its area, so larger islands are predicted to lose a smaller fraction of their species in a given period of time than a small island. For example, Trinidad, with an area of 4828 square kilometers, is predicted to have lost about 1 percent of its bird species diversity in the first century after isolation, whereas Isla Rey (an island of 249 square kilometers off the coast of Panama) is predicted to have lost nearly 4 percent of its birds in the same time.

Is 1 percent per century an acceptable rate of species loss? The answer depends on whether the lost species are biologically important and on whether extinction strikes at random or follows a consistent pattern. The accumulating evidence from a number of land-bridge island systems suggests that species tend to disappear in a strikingly consistent order, supporting the notion of a strongly deterministic extinction process. This consistency is bad news for conservationists, because it means that, following fragmentation of the landscape, the same species will be in trouble everywhere. The loss of the most extinction prone species might still be tolerable if these species played only minor roles in the ecosystem, but unfortunately the species that seem to disappear first are ones that appear to be indispensable in maintaining biological diversity—the top predators and nomadic frugivores.

The Biological Dynamics of Forest Fragments

Studies of land-bridge islands have projected the rates of extinction in isolated habitat fragments, but they reveal nothing about the biological mechanisms of extinction. Understanding these mechanisms is an urgent priority of conservation biology. Soon we shall no longer have the luxury of launching crash programs to rescue individual endangered species, whether they be tigers, pandas, or rhinoceroses. It will be necessary instead to scientifically manage whole ecosystems so that extinctions are held to a minimum. To achieve success, we will need a sophisticated knowledge of extinction mechanisms and of the species crucial to maintaining ecosystem stability.

Already it is clear that more than one mechanism is involved in the extinction process. Fragmentation of the forest produces extinctions directly when the resulting habitat patches are too small to meet the requirements of particular species. Some of the best documented examples are from Barro Colorado Island in Panama. BCI was created in the 1910s when the Chagres River was dammed to form Lake Gatun, which is now the central section of the Panama Canal.

Shortly after its creation, BCI became a favorite haunt of the eminent ornithologist Frank Chapman and a number of his colleagues from the American Museum of Natural History. The island thus came under the close scrutiny of some of the best naturalists of the day, and, thanks to this bit of serendipitous history, there is a well-documented record of the birds, mammals, reptiles, and amphibians that were present on BCI shortly after its isolation from the Panamanian mainland. We know, for example, that jaguars, pumas, harpy eagles, white-lipped peccaries, and great curassows frequented the island in the early 1920s, and that they all subsequently disappeared, presumably because the island was too small to support viable populations.

These frugivores and top predators have not been the only species to go extinct on BCI. After the early loss of species with the greatest space requirements, others have continued to disappear, including many originally common species that should have had no difficulty maintaining themselves. In 1970 another ornithologist, Edwin Willis, conducted a thorough resurvey of the birds of BCI. When he compared his results with Chapman's records of 50 years before, Willis found that a total of 45 species had disappeared over the five decades. Many of these species were birds of early successional habitats, whose loss would be expected as the island's vegetation matured. But more than a dozen were typical denizens of mature forest of the type that now occupies all but a tiny portion of the island. Why these species should have vanished remained a mystery.

Willis's meticulous records provide some clues. Because several of the extinct forest-dwelling species nest near or on the ground, Willis speculated that terrestrial mammals such

Barro Colorado Island sprawls like a giant green amoeba in Lake Gatun, Panama.

as coatimundis (a tropical raccoon) and opossums might be robbing nests with such frequency that the affected species could not reproduce. Support for this view was contained in the finding that one closely studied species failed in 96 percent of its nesting attempts because of predation.

Another clue is provided by Willis's population counts of two species that went extinct during the period of his observations. In neither case was there any evidence of the abrupt decline that would be expected in response to a sudden trauma such as a period of climatic stress. Instead, the two populations declined gradually but inexorably over a number of years, as the recruitment of new individuals consistently failed to compensate for mortality. Conditions on the island had somehow become inimical to these species. If so, extinction could not be regarded as a chance event, but rather it appeared to be an entirely predictable consequence of the prevailing conditions. It therefore remained to discover how conditions on the island might have changed after isolation in a way that led inexorably to the observed extinctions.

A Possible Role of Top Predators

As often happens in science, the answer came unexpectedly in the results of an unrelated study. The key research was conducted at the Cocha Cashu Biological Station on the Manu River in Amazonian Peru, one of the very few sites anywhere in the humid tropics to offer an environment entirely free of human influences. Rubber tappers had used the area as a thoroughfare in the early decades of the century,

The jaguar (*Panthera onca*) maintains the populations of many prey species at low levels, establishing a "balance of nature" in the Neotropical forest.

and in so doing had killed, enslaved, or driven away the indigenous inhabitants. After the establishment of rubber plantations in Asia rendered the gathering of natural rubber in Amazonia unprofitable, the Manu region was essentially abandoned, leaving a no-man's-land along the lower course of the river. The resulting wilderness offered an ideal setting for a research station that has been operating since 1970.

In the early 1980s, mammalogist Louise Emmons conducted a study of the top three terrestrial predators at this site—jaguar, puma, and ocelot. Using a variety of methods, she captured from one to several individuals of each species and fitted them with radio collars that enabled her to track their movements and habitat use. The predators' diets were reconstructed by identifying the undigested bones, scales, and hair in fecal remains. Emmons was able to collect large numbers of samples thanks to the habit all three cats had of walking the station's trails. The species that left each deposit could

be identified from the hair content, since cats swallow their own hair while grooming themselves. During the same period Emmons undertook exhaustive censuses of diurnal and nocturnal mammals to determine the availability of prey.

The jaguar proved to be an opportunistic generalist, taking not only mammals but also large numbers of reptiles (turtles and crocodilians) and occasional fish and birds. In contrast, the puma and ocelot proved to be dedicated mammal eaters. Since ocelots are the smallest of the three species, they mostly consumed prey weighing less than 1 kilogram, whereas the two larger cats generally took larger prey.

When Emmons considered the three predators and their mammalian prey together, an unexpected result emerged. Prey species appeared in the cats' diets at frequencies that almost perfectly matched their relative abundances in the environment. This correspondence between diet and availability of prey is evidence that the three cats are not selective in the prey they attack. Radio-tracking showed that they all hunted by walking slowly at a steady pace. In the close cover of a tropical forest, there is little opportunity to run down fleeing targets in hot pursuit. Prey are apparently either captured at the point of encounter or missed. There is no winnowing of herds to cull out the sick or the weak, as there is among lions and cheetahs on the plains of East Africa.

Because rain forest cats will attack any mammal they encounter, their abundance is largely determined by the prey species having the highest reproductive rates. Together only two species account for a large fraction of the production of prey biomass in the ecosystem: capybaras, which produce litters of up to four, and peccaries, which have litters of two or three that grow rapidly to large size. Species such as coatimundis, agoutis, and pacas produce only one or two young at a time and therefore contribute considerably less. If maintained at relatively high densities by the availability of rapidly reproducing prey species, large cats could reduce the densities of less fecund prey species.

That large predators exert an impact on less productive prey species is suggested in comparing the densities of terrestrial mammals at Cocha Cashu and at BCI. Several species, including coatimundi, agouti, and paca, are an order of magnitude more abundant where large felid predators are absent (BCI) than where these predators are present in natural numbers (Cocha Cashu). While there are perhaps other explanations for the observed differences in abundance, the most straightforward is that medium-sized terrestrial mammals have become superabundant on BCI because they live in an environment without predators. If this explanation is true, it follows that in the presence of predators and more productive prey, the densities of these animals are suppressed to levels far below those that could be sustained by the available food supply. If allowed to increase dramatically in the absence of predators, these same animals could begin to overharvest their food supplies and produce what have been termed "indirect effects."

This conjecture is especially plausible in the cases of the coatimundi, an avid predator of bird eggs, nestlings, and other small vertebrates; the agouti, a granivore that consumes the seeds of many species of mature-phase canopy trees; and the paca, a frugivore/granivore that consumes large numbers of recently germinated seedlings, as well as fruit and seeds. Coatis, for example, appear to be at least 20 times more abundant on BCI than at Cocha Cashu. It is thus not far-fetched to imagine that coatis at such an extraordinary density could be the agent that has caused the extinction of some of the birds on BCI.

This conclusion is supported by Willis's previously cited finding that the nests of certain bird species were subject to extremely high rates of predation. More recently, Willis's observations have been corroborated by direct experimental evidence. *Coturnix* quail eggs were set out in artificial nests on BCI and at a site on the nearby mainland where most of the extinct BCI birds were still common. Of 101 nests followed at the mainland site, only 4 were lost to predators, while on BCI the toll was 35 out of 51. It may not be fair to conclude that nest predation is more than fifteen times greater on BCI, but nevertheless nest predation on BCI is unusually frequent, and it is plausible that certain species have disappeared as a consequence.

A family of capybaras (*Hydrochaeris hydrochaeris*) lounges in a Venezuelan slough. The largest of all rodents, these fecund, 60-kilogram herbivores abound in South America's tropical wetlands, but in forested country they are largely confined to river banks.

As the table clearly shows, coatimundis are not the only prey of large cats to have become more abundant on BCI. Agoutis and pacas have similarly increased in number. Extraordinary numbers of these and perhaps other vertebrate seed predators could be suppressing the recruitment of certain tree species on BCI by consuming abnormally high numbers of seeds and seedlings.

Scientists working at BCI have recently begun to test this proposition. First, they have shown that seeds and seedlings of two large-seeded, mature-phase rain forest trees, *Dipteryx panamensis* and *Gustavia superba*, are up to 10 times more likely to survive on the Gigante Peninsula of the adjacent Panamanian mainland, where hunting keeps populations of mammals low. Second, a trio of BCI biologists have recently reported that the forests of several 70-year-old Lake Gatun islets are strikingly distinct from those on either BCI or the Panamanian mainland. On these islets, an absence of large seed predators could have led to reduced diversity of the tree community.

In particular, the tree community of several islands was dominated by two or more of the following large-seeded species: *Protium panamense* (Burseraceae), *Scheelea zonensis* (Palmae), *Oenocarpus mapora* (Palmae), and *Swartzia simplex* (Leguminosae). There is certainly more than one possible explanation for the highly unusual tree composition of these small island forests. Nevertheless, the findings are consistent with a scenario in which an absence of agoutis and other granivorous mammals has allowed a greatly increased recruitment of several tree species having large seeds that are known to be eaten by vertebrate seed predators. The Lake Gatun islets persuasively demonstrate the power of "indirect effects" to regulate the species diversity of tropical forests. In this case, a lack of top predators appears to lead to a superabundance of many prey species, and the abnormal numbers of these, may alter the pattern of tree recruitment. Such chain reactions may prove to be pervasive in ecology, and they may provide the key to maintaining biodiversity throughout the tropics.

Population Densities of Terrestrial Mammals at Cocha Cashu and Barro Colorado Island

	Number of individuals per square kilometer	
	Cocha Cashu	Barro Colorado Island
Didelphis marsupialis (opossum)	20	47
Dasypus sp. (armadillo)	4	53
Silvilagus brasiliensis (rabbit)	+	7
Dasyprocta sp. (agouti)	5	100
Agouti paca (paca)	4	40
Nasua narica (coati)	+	24
Tapirus sp. (tapir)	<0.5	0.5
Tayassu tajacu (peccary)	6	9
Mazama + Odocoileus (deer)	2.6	2.7

Overview

Top predators appear to regulate the abundance of many of their prey in the Neotropical forest, among them "mesopredators," such as the coati, and seed and seedling predators, such as the agouti and paca. In the absence of top predators, these animals increase dramatically in number and begin to exert indirect effects on the ecosystem: extinction of ground nesting birds and possibly other small vertebrates in the first case, and altered recruitment of tree species in the second. In these indirect effects one finds mechanisms that can account for the extinction of numerous species of vertebrates and plants. Furthermore, indirect effects may explain the loss of biodiversity that results from habitat fragmentation, as top predators are typically the first elements of the fauna to disappear in human-dominated ecosystems.

Although such a conclusion must be regarded as tentative until corroborated by additional studies, its implications for the future management of isolated forest parks are far-reaching. If indeed jaguars and pumas limit the numbers of many terrestrial mammals, and if some of these, in turn, are important predators of bird nests, small vertebrates, seeds, and seedlings, we shall have to recognize that, in order to preserve diversity in tropical forests, we will have to maintain a more or less natural balance between top predators and their prey. Disrupting this balance—by persecuting top predators; by overhunting pacas, agoutis, peccaries, and perhaps other game species; or by fragmenting the landscape into patches too small to sustain the whole interlocking system—could lead to a gradual and perhaps irreversible erosion of diversity at all levels.

Ironically, it emerges that species so rare that an astute observer is lucky to see one in a year of fieldwork—the tiger, the jaguar, the leopard—hold the key to ecosystem stability. If we unwittingly eliminate these species, as we have their counterparts over most of Europe and North America, human managers may attempt to emulate their function in an effort to preserve diversity, but it is doubtful we could ever do it so well.

9

Managing Tropical Forests

The deductions from island biogeography considered in the previous chapter portend a gloomy future. If, as seems reasonable, about 5 percent of the world's tropical forests are eventually brought under some kind of formal protection, the total area of habitat may not be adequate to ensure the perpetuation of diversity over the long run. Species are sure to disappear if, as has almost invariably been the practice, individual conservation units preserve no more than 1 percent of the original extent of a habitat. At best, remnants of this size will contain only a sample of the diversity present in the original landscape; at worst, they may not be able to sustain viable pop-

A herd of zebu cattle marches single file through the morning mists of Rondonia, Brazil. Government-subsidized programs to promote cattle ranching were the largest single cause of deforestation in the Neotropics in the 1980s.

ulations of key species such as top carnivores and migratory frugivores. The disappearance of large mammal species from major national parks in the United States and in East Africa, for example, has already been amply documented. Is then a cascade of extinctions in the tropics inevitable?

The answer depends entirely on how the rest of the landscape is managed. If wholesale deforestation—such as that now occurring in Brazil, West Africa, Indonesia, and elsewhere— continues unchecked, much diversity will almost certainly be lost. Alternatively, the cascade of extinctions could be prevented if large tracts of seminatural forest can be preserved outside of parks. However, there will be no incentive to keep large areas forested unless ways are found to make forests economically profitable without destroying them. For tropical nature to survive in recognizable form, tropical countries will have to bypass a stage in development passed through by virtually all industrial nations, namely, the systematic clearcutting of the virgin forest. A major portion of temperate diversity survived this insult; evidence presented in this and the last chapter suggests that tropical diversity will not fare so well.

More than 99 percent of the plant and animal species of North America and Europe survived the loss of the virgin forest. Few extinctions occurred, in part because deforestation took place over several centuries, so that some areas had partially recovered before others were affected. In contrast, deforestation in the tropics is proceeding so rapidly that whole nations will be denuded before there is any significant recovery. The forest will soon have vanished entirely in a number of countries, among them El

In the absence of ungulates or other large terrestrial mammals, cassowaries (*Casuarius casuarius*) fill an important role as seed dispersers in the forests of New Guinea, scavenging fallen fruit and passing the seeds through their guts.

Salvador, Haiti, Ivory Coast, Nigeria, Madagascar, and the Philippines.

Seedlings are able to invade new sites when birds and mammals transport seeds away from the parent tree. A majority of temperate trees grow from seeds that are dispersed by wind or small birds; these trees are thus not dependent on game birds or mammals for their reproduction. Those that are include oaks, hickories, chestnuts, beeches, walnuts, persimmons, and a few others, but these exceptions are adequately dispersed by squirrels, opossums, and other small mammals that are able to persist in proximity to man. In contrast, clearcutting of the tropical forest eliminates many birds and mammals that transport seeds, both by destroying their habitat and by encouraging hunting. In mainland tropical forests, the seeds of a sizeable majority of the trees are dispersed by large birds and mammals that are considered game by human populations. These species are systematically extirpated in the vicinity of human settlements. Major dispersers include such highly edible fare as ungulates (deer, antelope), primates (monkeys, apes, and relatives), the agouti and other large Neotropical rodents, guans, toucans, turacos, hornbills, and, in New Guinea, cassowaries.

Recovery from Disturbance in the Tropics

Extensive clearcutting of tropical forests, coupled with the elimination of seed dispersers, delivers a double blow to biodiversity that may require centuries to heal. To be sure, trees will spring up, and after 50 to 100 years the regrowth may resemble a normal tropical forest, but the successional forest will often be but a pale shadow of the original ecosystem. Plant diversity will be greatly reduced, and many birds, mammals, reptiles, and other, lesser creatures will not quickly return.

This inference can be drawn from two kinds of observations, one anecdotal and one based on systematic measurements. The anecdotal evidence comes from personal observations at Tikal in Guatemala, a site that was abandoned centuries ago after serving as a center of the Maya civilization.

Archaeological remains suggest that hundreds of square kilometers surrounding the great pyramids at Tikal were cleared and intensively cultivated during the Maya Classic Period, which lasted from A.D. 300 to 900. Mysteriously, this great civilization came to a precipitous end in the ninth century A.D., and the district was largely abandoned. Over the ensuing 1200 years the forest at Tikal has recovered, even attaining grand proportions, and wildlife now abounds.

Nevertheless, two features of the forest at Tikal will catch the discriminating botanical eye: first, the plant diversity is anomalously low, and second, many of the common tree species are ones known to have been used by the Maya, either because they possessed desirable wood or because they bore edible fruits or nuts. Extraordinary concentrations of the latter undoubtedly contribute to the high densities of peccaries, monkeys, turkeys, parrots, and other wildlife to be seen today in the ruins. Here, in a situation that would be unusual in contemporary practice, humans partially substituted for the missing animal dispersers. Still, after 1200 years, the site has not recovered a normal level of plant diversity.

That diversity recovers very slowly after major disturbance is well illustrated in a study undertaken by the author and several colleagues in an Amazonian floodplain. We took advantage of the active meandering of the Manu River to study the process of recovery. Long tongues of land are created within each meander loop as the river progressively cuts away at

Abandoned more than 1000 years ago, this Mayan temple has been reclaimed by the forest. Even after a millennium, plant diversity at this highly disturbed site may not have fully recovered.

the outside bank during annual floods. Where the banks are low, or composed of sandy substrate, the annual advances can be quite impressive, even 50 meters or more. While this process is extending the outer perimeter of bends, fresh sediment is being added to their inner margins. When the river recedes at the end of the wet season, the sediment is exposed in the form of long beaches, which are promptly colonized by plants of many species. Growth is especially rapid because the fresh alluvial deposit is extremely fertile, having been recently eroded from bedrock on the Andean slopes.

Among the plants that sprout on the open beaches are seedlings of a number of "pioneer" trees—species that constitute the vanguard of a prolonged successional process. Some of these grow very quickly, reaching heights of 8 to 10 meters in just three years, while others develop more slowly. The fast-growing species never attain great stature, and most of them are short-lived, so that after a decade most of the pioneers have died and been replaced by slower growing but taller and longer-lived species. Still other species gradually grow up under these, eventually overtopping them to assume dominant positions in the canopy. The forest goes through several generations of progressive change in composition before it approaches the mature state after perhaps 300 years. Throughout the prolonged period of successional change, the stature, species diversity, and vertical complexity of the forest continue to increase. The diversity and abundance of wildlife increase in parallel, attaining their highest levels only in the most mature phases.

Forest that is 100 to 150 years old has only accumulated about half the birds, mammals, and trees of the mature forest. Many of the tree species present in a forest of this age are characteristic of early successional stages.

For example, in one 0.5-hectare plot estimated to be about 100 years old, there were 164 trees of 49 species. Of the 164 individual trees, 111, or 67 percent, were early successional species, including 12 of the 15 most abundant. Most species of the mature forest were represented by only one or two individuals. Even after a century, the eventual tree community had only just begun to colonize the site.

The tip of this Amazonian meander loop gradually elongates as the river eats away at the opposite bank during annual floods. The new land is quickly claimed by vegetation, initiating a plant succession that must continue for centuries before plant and animal diversity attain their highest levels.

Diversity at Several Stages of Succession in an Amazonian Floodplain

	Number of species			
	Pioneer stage (3 to 5 years)	Early succession (30 to 50 years)	Late succession (100 to 150 years)	Mature forest (>300 years)
Birds	21	49	127	236
Primates	0	2–6	6–8	8–12
Trees*	19	33	50	112

*Number of species with a diameter of at least 10 centimeters at breast height contained in 0.5-hectare sample plots.

Elizabeth Losos, a Princeton University graduate student, found that very few seeds of mature-phase species enter early successional zones, and that most of these are brought in by birds. By sowing seeds and transplanting seedlings, she showed experimentally that some mature-phase species can germinate, survive through the seedling stage, and grow rapidly under early successional conditions. Her results support the conclusion that it is the slow dispersal of seeds into the early stages that limits the pace at which diversity recovers after large-scale disturbance.

Losos did her research in a near-optimal environment where mature forest was present within 500 meters, where hunting was prohibited, and where all animal dispersers were present at natural densities. None of these conditions pertain to the huge man-made clearings that have become commonplace throughout the tropics. For diversity to recover in such sites as these, as at Tikal, 1200 years may not be enough.

A researcher provides the scale for judging the development of a 20-year-old secondary forest in Venezuela. Such a forest might have accumulated enough nutrients to serve in a slash-and-burn rotation, but will require many more decades to recover its timber volume.

As a footnote to this discussion, I should mention that the criteria by which a timber operator might judge the quality of a stand (tree girth, basal area, cubic meters of wood volume per hectare) reach maximal or near maximal values in perhaps 100 years, less than a third of the time the forest requires to attain the highest diversity of birds, primates, and trees. The early appearance of desirable timber trees does not bode well for the retention of high levels of biodiversity in intensively exploited tropical forests.

The Response of Tropical Wildlife to Timber Harvest

Timber harvest invariably alters the habitat, even when relatively few trees are removed. In addition to reducing the density of trees, logging alters the composition of the forest and often lowers species diversity. Whereas a primary forest may contain a patchwork of early successional openings embedded in a matrix of mature canopy, a harvested forest will often contain remnant groves of mature trees in a matrix of early successional regrowth.

Within the last decade a number of studies have examined the impact of logging on wildlife, particularly on primates and birds. These studies have compared animal populations in undisturbed primary forests with those in selectively logged forests in Amazonia, Africa, and Southeast Asia. (In selective logging, timber operators harvest only valuable species.) Despite geographical differences in flora and fauna, the results of these studies are remarkably similar. Frugivorous primates and birds, particularly the larger ones, decline in number after disruption of the canopy. Folivores, on the other hand, are often unaffected. Indeed, their numbers may even increase as the regenerating vegetation offers a greater abundance of young leaves.

Although the mechanisms causing the decline of frugivores have not been identified conclusively, biologists have presumed that a loss of important food resources is a major factor. Even highly selective logging can severely damage a forest. An exemplary study by British forester Andrew Johns, for example, observed the results of harvesting only 3 percent of the trees in a West Malaysian dipterocarp forest. He found that, despite the low percentage of trees harvested, 51 percent of the stand was destroyed. In addition to the 3 percent of trees that were extracted, another 5 percent were pushed aside in road building, and 43 percent were injured or killed during felling and dragging operations. The damage extended over all species and sizes of trees. In this case the harvested trees belonged to a family (Dipterocarpaceae) that does not play an important role in primate nutrition. Nevertheless, the loss of other trees lowered the production of fruit, while young foliage remained abundant.

In other regions, the preferred timber trees may be an important source of food for wildlife, and the removal of these species can precipitate the abrupt disappearance of certain primates. Efforts to enhance the commercial value of forests can jeopardize other wildlife as well. For example, the poisoning of commercially valueless fig species destroys an essential component in the diet of many rain forest birds and mammals.

In one of the few studies to compare different levels of logging activity in the same region, University of California graduate student John Skorupa found in Uganda that after more than 50 percent of the canopy was destroyed, the numbers of primates were severely reduced,

BEFORE

— 40 meters

— 20 meters

— 0

AFTER

— 20 meters

— 0

whereas less drastic harvests had relatively mild effects. It seems likely that the duration of adverse conditions will depend on the extent of damage to the canopy. Where more than 50 percent of the canopy is removed or damaged, the negative impact may persist for a number of years; where less than 50 percent has been damaged, habitat quality can be restored relatively quickly because the remaining trees rapidly expand into the gaps. Such results argue for the adoption of logging methods designed to minimize physical damage to the forest.

Extractive Reserves as a Model of Sustainable Management

Commercial timbering, especially when carried out with heavy machinery, does long-lasting damage to the forest and the soil beneath it. Conservationists are seeking less harmful ways of deriving economic benefits from tropical forests. Ecotourism is one form of use that is currently being promoted. Another is to harvest nontimber products. In many parts of the world local peoples have long taken products such as fruits and medicinal herbs from tropical forests without seriously jeopardizing biodiversity. Included in the long list of such products are a number that enter international commerce, such

(Opposite page) Extraction of only the largest 3 percent of the trees in this Malaysian forest leaves a wake of destruction. Careless felling and the intrusion of heavy machinery needlessly kill or damage half the trees, greatly diminishing the economic value of the residual forest.

as natural rubber, chicle, Brazil nuts, rattan, fruits, gums, latex, fiber, medicinal plants, game, and live animals for zoos and medical research. From a conservation standpoint, these traditional, village-based extractive activities represent an ideal form of utilization because the forest remains essentially intact.

The world's attention was focused on the nondestructive use of tropical forests by the 1989 murder of Chico Mendes, leader of the Brazilian rubber tappers' union. The union had come into conflict with cattle ranchers who wanted to create pasture out of forest that had been providing the rubber tappers with a livelihood for generations. The resulting clash of values is typical of a process that has been quietly taking place all over the world as powerful interests have displaced politically impotent people from the land they have used for decades or centuries.

The murder of Chico Mendes exposed the conflict over land rights in Brazil's state of Acre to intense international scrutiny. Embarrassed by the unfavorable publicity, the Brazilian government agreed to the formal establishment of an "extractive reserve" in Acre where the lifestyle and livelihood of the rubber tappers would be secure.

Now that the emotion has subsided, we must ask whether such extractive reserves truly make good economic sense. The move to create extractive reserves in Brazil and elsewhere in the tropics has been propelled by a number of highly optimistic economic analyses. These reports show that revenues from the sale of natural forest products can potentially exceed the income from more destructive forms of development, such as timbering and cattle ranching.

A Guatemalan "chiclero" carefully incises the bark of a chicle tree (*Manilkara zapota*) to extract the valuable latex used in chewing gum. Traditional extractive economies such as this have sustained rural people for generations, yet do not seriously compromise the forest's biodiversity.

Let us consider two analyses from Amazonia, the region I know best.

A much publicized article published in 1989 by a distinguished multidisciplinary team composed of a botanist, an ecologist, and an economist evaluated the annual production of fruits, nuts, gums, and timber produced by a single hectare of forest in the vicinity of Iquitos, Peru. These products were assigned values based on current retail prices in the Iquitos market. The authors calculated a predicted gross revenue of $650 per hectare per year and estimated the net present value of the standing forest at $8890 per hectare. These amounts compare extremely favorably to those obtained from alternative forms of land use in the same region, such as pasture and pulpwood plantation, but the numbers represent theoretical conjecture, not actual practice.

In another article published in 1989, anthropologist Stephan Schwartzman described the income to families dependent on a mixed economy of rubber tapping and Brazil nut gathering in the state of Acre. An average family produces about 750 kilograms of rubber and 4500 kilograms of Brazil nuts a year from a holding of some 200 hectares. The families cultivate gardens and raise livestock as well, but only for their own subsistence. At prevailing prices, cash earned from selling the rubber and Brazil nuts generates an annual income of $960 per family. For each hectare of forest, the return is $4.80 per year. This actual return can be contrasted with the hypothetical return of $650 per year stated in the preceding paragraph.

It is likely that hidden assumptions in the first example can account for much of the discrepancy between the two figures. Embedded in the economic arguments are assumptions about prices, markets, and other relevant conditions that crumble when examined critically. Conservationists, in the desperation of their predicament, have a tendency to grasp at straws. In the favorable economic analysis, wishful thinking has undermined objectivity.

Heedless of the risks to fingers and toes, this "castaneiro" opens coconut-sized Brazil nut fruits to extract the familiar nuts. Although environmentally benign, such rudimentary technologies limit workers to an economically marginal existence. The spread of plantations and mechanization seem likely to eliminate this traditional lifestyle within a generation.

A more realistic view is that prices in the Iquitos market represent a balance between supply and demand. Given that Iquitos is surrounded by Amazonian forest, the supply of forest products is virtually unlimited; it is therefore the demand that determines the price. In order for more people to obtain incomes by gathering fruits and nuts, supply would have to be increased without seriously depressing prices, a highly unlikely prospect. Most minor nontimber products, such as fruits, nuts, gums, latex, and medicinal plants, are unknown outside the countries of origin, and within these countries have extremely limited markets, easily depressed by oversupply.

Some well-intentioned organizations are attempting to expand demand by promoting export markets for Amazonian forest products. My prediction is that these efforts, if successful, will have the unintended result of increasing deforestation. Levels of production in natural forest are low, and the harvest is typically of poor quality. For example, wild fruit is often attacked by fungus or insect larvae, and frequently suffers bruises in falling from the high canopy. Their poor quality renders such products unsuitable for international commerce.

Moreover, it is doubtful that natural forests will be able to compete with plantations as a source of any plant product that is marketed in quantity. A case in point is natural rubber. The continued harvest of rubber in Brazil rests on extremely tenuous economic conditions, as revealed in a passage by Philip Fearnside, one of

Road construction in the Brazilian Amazon. Large sections of the 32,000-kilometer Transamazonian highway system have washed out in the torrential rains and are passable only in the dry season or else have been abandoned altogether.

the foremost experts on Amazonian development. "A key factor in making the extractive reserve scheme viable is the price of rubber. Rubber in Brazil is heavily subsidized by government pricing policies. Because *Microcyclus* fungus does not exist in Southeast Asia, plantation rubber is inherently cheaper to produce there than it is in Amazonia. World rubber markets have been depressed in the 1980s to the point where many productive plantations in Indonesia and Malaysia have been cut to replant with other crops. Brazil imports two-thirds of its rubber; the remaining third is produced within the country and bought at a price that, although low from the point of view of rubber tappers, is far above that of international commodity markets. The difference represents a subsidy that is being paid by Brazilian consumers when buying products made of rubber."

Today, there is a strong global trend toward free trade and the elimination of artificial markets. In light of this trend, the future of extractive reserves based on natural rubber appears highly problematical. Any natural product that does enjoy a strong international market will be produced in plantations, current technology permitting. Such plantations are likely to be established at the expense of natural forest, hence the prediction that promotion of markets for natural forest products will lead to further deforestation.

All actual and proposed extractive reserves are located in sparsely populated regions where economic development is hindered by a lack of transportation. Penetration of these areas by roads will be the death knell of extractive reserves, for governments around the world have proven powerless to stanch the tide of land-hungry subsistence farmers. Even where protected areas were funded by the World Bank, as in Rondonia, all were invaded by loggers, miners, or colonists as soon as roads provided access.

Fundamentally, gathering is an economically marginal activity. Even in portions of Acre where natural rubber and Brazil nut trees attain their highest natural densities, tapper/collector

A glimpse of the future? When cleared, the land occupied by seasonally flooded varzea forest has proven suitable for intensive rice cultivation. Modern agriculture has not spread widely in Amazonia because villagers lack the capital to invest in the necessary dikes and water control systems.

families require 200 hectares to meet their needs. To justify allocating land to extractive reserves, governments would have to be persuaded that there were not available more productive and politically attractive forms of land use. In some remote roadless areas conditions might favor extractive reserves for a time, but more intensive land use seems inevitable wherever population pressure begins to build. Instead of the 200 hectares needed for a family of "shiringeiros," 50 hectares suffices for a family of slash-and-burn agriculturists. Even smaller plots can be adequate if managed with the technology of low-input cropping (recall Chapter 2). Will political realities permit low-intensity land uses in countries burdened with population densities of 50, 100, or 150 persons per square kilometer? It hardly seems likely.

What seems more probable is that intensive rather than extensive forms of land use will soon spread over most of the remaining tropical wilderness. Agricultural products and wood are basic to every economy in the world, and are required in ever greater amounts to meet the demands of population growth. Land that is even marginally suitable for farming will be cleared and settled. As forests continue to shrink, tropical hardwood will become more valuable. Meanwhile, the world market for wood is projected to expand into the foreseeable future. At some point wood will become so scarce and so valuable that forests will be able to compete economically with other forms of land use. What we must fear is that such a balance will not be achieved in the tropics until the primary forest has all but disappeared.

Conserving Biodiversity in Managed Tropical Forests

It seems clear that if forests are to continue to exist in the tropics, they will have to justify themselves economically. Forests will have to support remunerative activities in addition to the extraction of nontimber products. In recognition of this necessity, conservationists have recently begun to promote what is known as "natural forest management" (NFM). The details of implementation can vary, but the goal of NFM is the extraction of timber from stands of native trees composed of a diversity of species. Its practitioners design ways to manage the forest that will increase the productivity of commercially valuable timber species. Tree species composition will change as an inevitable consequence of almost any form of management, so forests under NFM do not always retain their original character.

Natural forest management is not a new concept; it is an old concept undergoing revival. Before many Old World tropical countries were granted independence around 1960, colonial administrations had investigated NFM and put it into wide practice. Many of the original management schemes failed to enhance timber production as much as desired, and they were abandoned in disillusionment. The concept has now returned to the forefront of tropical forestry in a new guise—no longer is NFM satisfied merely to enhance timber revenues. Proponents of the new NFM advocate that management plans specifically take into account the perpetuation of biodiversity. In practice, preserving biodiversity will require managing

tropical forests not only for their trees, but for their animal populations as well, so that important seed dispersers do not disappear.

As I write in 1991, NFM in the tropics is more a dream than a reality. The International Tropical Timber Organization estimates that less than 1 percent of tropical forests worldwide are being sustainably managed. Primary tropical forests are now typically "high-graded" (selectively cut) for the most valuable timber species, and then abandoned without replanting. In many parts of the world, the abandoned land is soon invaded by shifting cultivators, so that second and subsequent timber harvests are never realized. In some cases, large tracts of forest have been cleared to make way for tree plantations of alien species. There is no denying that plantations of pine or eucalyptus can often produce wood at a higher rate than a natural forest composed of hundreds of species of trees. However, when forest lands are dedicated to the sole purpose of maximizing wood production, the benefits of a major public resource are denied to any constituency other than the timber industry.

The inherent inequity in such a one-sided approach to resource management was recognized more than 80 years ago by Gifford Pinchot, the founding father of the U.S. Forest Service. Pinchot, a man of resolute character, perceived that public forests have multiple constituencies, and that management should attempt to balance the objectives of many competing and often contentious interest groups. The potential benefits of forests are numerous. In addition to supplying timber, forests provide other, less quantifiable assets such as watershed protection; public access to hunting, fishing,

A worker cuts a *Manilkara* log into lengths to extract commercial timber from a Brazilian forest before the land is totally cleared for replanting as a tree farm. Experience with plantation forestry in the humid tropics has been mixed. The trees are often attacked by pests and diseases, and growth on infertile soils can be disappointingly slow.

and recreation; and a source of nontimber products. Last, but not least, forests serve as reservoirs of biodiversity.

Is natural forest management compatible with the long-term maintenance of biodiversity? The answer is yes, in principle. Several scientifically based systems have been instituted in different parts of the world by knowledgeable tropical foresters. Since the systems are conceptually quite diverse, I shall briefly describe four of them. Although only one included the perpetuation of biodiversity as an expressed goal, they all share the common ideal of sustainability, at least with respect to timber production.

Devised by British foresters under the pre–World War II colonial administration, the Malaysian Uniform System (MUS) was long held up as the shining example of how scientific forestry could triumph in the daunting task of profitably managing stands containing several hundred tree species. The system is applicable only to Southeast Asia, however, because it takes explicit advantage of the synchronous, or "mast," fruiting of Dipterocarpaceae, the plant family that includes many of Asia's most valuable timber species, so-called Philippine mahogany.

As the reader will recall from Chapter 7, dipterocarps of many species fruit synchronously at irregular multiyear intervals. Following one of these masting episodes, multitudes of dipterocarp seedlings appear on the forest floor. The seedlings invariably die from a lack of light unless a gap appears in the canopy. Management consists of harvesting the overstory while there are seedlings and saplings present that can take advantage of the increased light. Dipterocarps thus give rise to more dipterocarps. The MUS is a "uniform" system because each stand is harvested as a crop. Outside of Southeast Asia, such uniform systems have not succeeded because the commercially valuable species of the mature forest are not well represented in the regrowth.

The tropical shelterwood system is another management scheme devised by British foresters during the colonial period. It was practiced extensively in West Africa, where the most desirable species were found to regenerate best in semishaded conditions. Foresters employed a variety of techniques to enhance the quality and productivity of the regeneration. Vines were cut, the understory was cleared to encourage the growth of seedlings, and noncommercial species were girdled or poisoned to allow more light to pass through the canopy. In another scheme, termed "enrichment planting," saplings of desirable species were planted in narrow strips cleared in the forest at 20-meter intervals. The standing forest between the strips provided the required partial shade. Unfortunately, neither of the above management systems, nor several others that were tried concurrently, gave encouraging results. Recruitment of desirable species in the regrowth was variable and unpredictable, being nearly always far below expectation, and growth rates were slow.

The practice of these systems of "natural management" in West Africa has now been largely abandoned.

The CELOS system is a much more recently devised scheme developed by Dutch foresters for managing New World tropical forests in Suriname. In contrast to the Malaysian Uniform System, the CELOS system is polycyclic; that is, only a few trees at a time are removed from a hectare, and harvests are conducted periodically at intervals much shorter than the time required for an individual tree to grow to maturity. Another important feature of the CELOS system is that 40 to 50 species are managed, not just the one or two having the highest value. Much care is taken during harvest to lessen the damage to the residual forest and the soil. Most of the large noncommercial trees are killed by poisoning. The standing volume is thereby reduced to about half, and the forest is opened to light, which stimulates the growth of undersized individuals of commercial species. Under this treatment, the annual increase in the volume of commercial timber has grown by a factor of about four, a heartening result in view of the disappointing experience with similar techniques in West Africa.

The fourth management system we consider was devised by Gary Hartshorn, a U.S. forester who, with USAID sponsorship, established a cooperative forestry enterprise in the Yanesha Amerindian community in the Palcazu Valley of Peru. Hartshorn's system is designed to retain diversity by encouraging regeneration from seed in long, narrow clearcuts. The use of clearcuts permits the harvesting of much larger volumes of wood per unit area than are extracted under the high-grading practiced so

In time, nature heals its wounds. Attractively forested now, this landscape in the valley and ridge province of central Pennsylvania was once a charred wasteland. Massive wildfires burned over much of the state after careless and wasteful logging of the virgin forest in the nineteenth century.

widely in the tropics. Up to 350 cubic meters can be harvested from a hectare: 150 cubic meters as saw logs, 90 cubic meters of round-wood for poles and posts, and the remainder as branchwood sawn for specialty items or converted to charcoal. The advantage of using all the wood is that the area harvested can be greatly reduced.

The distinctive feature of the scheme is that the clearcuts are confined to narrow strips, 30 to 40 meters wide and 200 to 500 meters in length. In principle, seeds for the next crop of trees could freely enter the strips from the flanking forest. Indeed, crowns of the intact forest literally tower over the harvested strips. However, David Gorchov of Miami University

of Ohio has shown that surprisingly few seeds are dispersed into the strips. He captured only a tenth as many seeds in traps placed 12.5 meters into the strips as at the forest edge. Nevertheless, a survey of one strip 27 months after harvest revealed that 155 regenerating species represented by saplings at least 1 meter tall were present in only 0.15 hectare, more than double the number of species harvested from the same area. Once the regenerating stand has formed a closed canopy, noncommercial species are selectively removed. The system has the potential to maintain a very high level of plant diversity and thereby provide an adequate habitat for birds, primates, and other fauna.

Although a great deal of planning, scientific expertise, and effort of execution has gone into developing each of these forest management schemes, essentially all of them have failed. They have failed in some cases (West Africa) because of technical insufficiencies, but, more fundamentally, they have failed because social, political, and economic conditions in the countries involved have changed beyond recognition, disrupting the stability that is essential to any long-continuing venture.

In Malaysia, it was discovered that plantation culture of rubber trees and oil palms provided much higher returns than forest management, and so nearly the entire lowland forest of peninsular Malaysia has been cleared for replanting in these crops.

In West Africa, a demographic tidal wave has overwhelmed all but remnants of a once ample forest estate, as slash-and-burn agriculturists push into every remaining patch of unoccupied land. In the words of Chelunor Nwoboshi, a West African forester, "Under the prevailing circumstances, forestry needs evidence of impressive socioeconomic returns from forest lands to ward off the competition from other land uses. For example, only recently Nigeria lost about 10,000 hectares of forests in the Okomu reserve—one of the centers of natural forest management—to oil palm cultivation, and an estimated additional 280,000 hectares of productive forests were lost throughout the country."

In Suriname, a democratically elected government was overthrown in a bloody military coup, and most foreign interests were subsequently expelled from the country, thus terminating the CELOS experiment. And in Peru,

Shining Path guerrillas have imposed a reign of terror on the Palcazu Valley, forcing the USAID mission to abandon its operation there.

These are the realities of the tropical world that conservation efforts will have to cope with if they are to be successful. The failure of all four of these well-intentioned efforts to survive even one harvest cycle suggests that existing social and political conditions in many tropical countries are inimical to long-range planning and the sustainable use of forests. Nevertheless, if there is to be any hope for the survival of tropical nature, long-range planning and sustainable management must lie at the heart of it. Science can provide technologies for the rational and sustainable use of renewable natural resources, but politics and social action must bring about the necessary transformation of attitudes. Given the extremely rapid pace of change in the developing countries, and the intractable poverty and ignorance in which much of their populations live, one can seriously doubt whether these attitudinal changes will emerge in time to save the tropical forest.

It is my view that significant areas of tropical forest will survive into the middle of the next century only if the developed countries take the lead in preserving them. The influence of the developed countries in the poor nations of the tropics is far greater than most Westerners imagine. Almost nothing progressive happens in many low-income countries that isn't financed from Europe, North America, or Japan. Financially strapped Third World governments are incapable of investing in the future because their resources are fully committed in paying the salaries of public sector employees.

The mountainous Monteverde cloud forest of Costa Rica is typical of the natural forests that will remain in the humid tropics after another few decades. Because most forests will be preserved on steep slopes for watershed protection, lowland flora and fauna are in serious jeopardy in many parts of the tropics.

The developed countries can play a decisive role in preserving tropical forests, but only if they make the cause of perpetuating biodiversity a matter of high policy. Tropical countries could be encouraged to conserve forests through a push-pull system of incentives and disincentives. A government that took positive steps such as implementing policy reforms, instituting natural forest management, enforcing conservation legislation, or reducing population growth might be rewarded by direct financial assistance or it could be rewarded through more creative measures such as debt relief, reduction of trade barriers, or the provision of technical assistance. Countries that faltered in political will, or in the implementation of conservation policies, could face the loss of financial assistance, the canceling of trade agreements, or the closing of immigration quotas. To be effective, such a push-pull system would have to be backed by a political commitment on the part of the developed world that has so far not been evident.

The managed tropical forests of the future will generally be confined to the poorest soils and steepest slopes, where not even the most rudimentary forms of agriculture will be economically competitive. Nevertheless, forests managed for multiple uses on otherwise marginal lands can still be of vital service in supplying timber and nontimber products, fuel wood, fish, game, recreation, watershed protection, climate amelioration, and the perpetuation of biodiversity. The loss of intangible benefits such as watershed protection and climate amelioration is rarely figured as a cost in land use decisions. Multiple-use management and proper cost/benefit accounting can make forests an economically viable option for the future, even in the most famished lands. But to save biodiversity, we must act before the virgin forest disappears, because no effort at ecosystem rehabilitation, however sophisticated, will ever recreate nature in its primeval state.

Sources of Illustrations

Page 1
Michael and Patricia Fogden
Page 2
Chip Clark
Page 4
Michael and Patricia Fogden
Page 5
Adapted from C. H. Dodson and A. H. Gentry, "Flora of the Rio Palenque Science Center," *Selbyana* 4 (1978)
Page 7
Merlin D. Tuttle/Bat Conservation International
Page 9
André Bärtschi
Page 11
George Bernard/Natural History Photo Agency
Page 12
Based on a map published by the Smithsonian Institution.
Pages 13 and 14
André Bärtschi
Page 15
Adapted from H. Walter, *Vegetation of the Earth and Ecological Systems of the Geo-biosphere*, Springer-Verlag, New York, 1985
Page 16
Adapted from H. Walter, *Vegetation of the Earth and Ecological Systems of the Geo-biosphere*, Springer-Verlag, New York, 1985
Pages 17, 19 and 21
André Bärtschi
Page 24
Adapted from E. Pianka, *Evolutionary Ecology*, Harper and Row, New York, 1983
Page 26
Loren McIntyre

Pages 27 and 30
Photo by Gerald Cubitt. From *Wild Malaysia*, by Gerald Cubitt and Junaidi Payne. © New Holland Publishers, Ltd., London.
Page 35
Donald Perry
Page 37
Adapted from C. F. Jordan, *Nutrient Cycling in Tropical Forest Ecosystems*, Wiley, New York, 1985
Page 38
Adapted from T. C. Whitmore, *Tropical Rainforests of the Far East*, Clarendon, Oxford, 1984
Page 39
I. Polunin/Natural History Photo Agency
Page 41
Gerald Cubitt/Bruce Coleman Ltd.
Page 43
Adapted from C. F. Jordan, *Nutrient Cycling in Tropical Forest Ecosystems*, Wiley, New York, 1985
Page 44
Adapted from Heinrich Walter, *Vegetation of the Earth and Ecological Systems of the Geo-biosphere*, Springer-Verlag, New York, 1985
Page 45
Hanbury-Teison/Robert Harding Picture Library
Page 47
André Bärtschi
Page 49
Adapted from C. F. Jordan, *Nutrient Cycling in Tropical Forest Ecosystems*, Wiley, New York, 1985
Page 50
Pedro Sanchez

Page 51
Adapted from P. A. Sanchez et al., *Science* 216 (1982)
Pages 52 and 53
André Bärtschi
Page 54
Adapted from R. H. MacArthur, *Geographical Ecology: Patterns in the Distribution of Species*, Harper and Row, New York, 1972 [From MacArthur 1969]
Page 55
Photo by Gerald Cubitt. From *Wild Malaysia*, by Gerald Cubitt and Junaidi Payne. © New Holland Publishers, Ltd., London.
Page 56
Michael and Patricia Fogden
Page 60
Michael and Patricia Fogden
Page 62
Photo by Gerald Cubitt. From *Wild Malaysia*, by Gerald Cubitt and Junaidi Payne. © New Holland Publishers, Ltd., London.
Page 64
Adapted from J. Terborgh, *Acta XVII Congressus Internationalis Ornithologici*, Deutschen Ornithologen-Gesellschaft, Berlin, 1980
Pages 65, 66 and 67
Michael and Patricia Fogden
Page 68
Adapted from T. W. Schoener, *Condor* 75 (1971)
Page 71
Adapted from R. H. MacArthur, *Geological Ecology: Patterns in the Distribution of Species*, Harper and Row, New York, 1972
Pages 72 and 73
Loren McIntyre

Page 75
Adapted from A. H. Gentry, *Proceedings of the National Academy of Science*, U.S.A. 85 (1988)

Page 78
Adapted from P. Ashton, unpublished data

Page 80
Adapted from A. H. Gentry, *Annals of the Missouri Botanical Garden* 75 (1988)

Page 82
Larry Ulrich

Page 83
Marcos A. Guerra/Smithsonian Tropical Research Institute

Page 86
Adapted from S. P. Hubbell, *Science* 203 (1979)

Page 87
André Bärtschi

Page 88
Adapted from S. P. Hubbell, *Science* 203 (1979)

Page 89
Photo by Gerald Cubitt. From *Wild Malaysia*, by Gerald Cubitt and Junaidi Payne. © New Holland Publishers, Ltd., London.

Page 91
Linda Sims

Page 92
André Bärtschi

Page 94
Adapted from S. P. Hubbell and R. B. Foster, *Conservation Biology: The Science of Scarcity and Diversity* (ed. M. E. Soule), Sinauer, Sunderland, Mass.

Page 97
Phillip Rosenberg

Page 98
Michael and Patricia Fogden

Page 99
Michael Terborgh

Page 102
Jany Sauvanet/Natural History Photo Agency

Pages 104 and 105
George Bernard/Natural History Photo Agency

Pages 106 and 107
Adapted from T. C. Whitmore, *Tropical Rainforests of the Far East*, Clarendon, Oxford, 1984

Page 108
Michael and Patricia Fogden

Page 110
Adapted from J. Terborgh and K. Petren, *Habitat Structure: The Physical Arrangement of Objects in Space* (eds. S. S. Bell, E. D. McCoy and H. R. Mushinsky), Chapman and Hall, London, 1991

Page 112
Francis Hallé, *Institut Botanique*

Page 113
Adapted from J. Terborgh, *Acta XVII Congressus Internationalis Ornithologici*, Deutschen Ornithologen-Gesellschaft, Berlin, 1980

Page 114
Adapted from F. G. Stiles, *Science* 198 (1977)

Page 115
Michael and Patricia Fogden

Page 116
Adapted from T. C. Whitmore, *Tropical Rainforests of the Far East*, Clarendon, Oxford, 1984

Page 118
John Terborgh

Pages 119, 120 and 123
Adapted from J. Terborgh, *American Naturalist* 126 (1985)

Page 124
Adapted from J. Terborgh and K. Petren, *Habitat Structure: The Physical Arrangement of Objects in Space* (eds. S. S. Bell, E. D. McCoy and H. R. Mushinsky), Chapman and Hall, London, 1991

Page 126
André Bärtschi

Page 127
George Bernard/Natural History Photo Agency

Page 128
Donald Perry

Pages 130 and 131
Michael and Patricia Fogden

Page 133
Adapted from S. D. Webb, *Organization of Communities Past and Present* (eds. J. H. R. Gee and P. S. Giller), Blackwell, London, 1987

Page 134
Michael and Patricia Fogden

Page 137
Field Museum of Natural History (neg# GE084608), Chicago

Page 138
Michael and Patricia Fogden

Page 140
Illustrations drawn by Marlene Hill Werner for Dr. Larry Marshall, courtesy of Field Museum of Natural History

Pages 142 and 143
Adapted from J. Haffer, *Science* 165 (1969)

Page 145
Adapted from J. Turner, *Natural History* 84 (1975)

Page 146
Paul Colinvaux

Page 148
Frans Lanting/Minden Pictures

Page 149
Adapted from M. E. Soule, *Conservation Biology: An Evolutionary-Ecological Approach*, (eds. M. E.

Soule and B. A. Wilcox), Sinauer, Sunderland, Massachusetts, 1980
Page 150
Adapted from J. Terborgh, *American Naturalist* 107 (1973)
Pages 152 and 153
André Bärtschi
Page 155
Photo by Gerald Cubitt. From *Wild Malaysia*, by Gerald Cubitt and Junaidi Payne. © New Holland Publishers, Ltd., London.
Page 156
André Bärtschi
Page 157
Adapted from F. Bourliere, *Tropical Rainforest Ecosystems in Africa and South America* (eds. B. J. Meggars, E. S. Ayensu and W. D. Duckworth), Smithsonian, Washington, D. C., 1973
Page 159
C. B. Cox and P. D. Moore, *Biogeography: An Ecological and Evolutionary Approach*, 4th ed., Blackwell, London, 1985
Page 161 left
Michael and Patricia Fogden
Page 161 right
Morten Strange/Natural History Photo Agency
Page 163
E. A. James/Natural History Photo Agency
Page 164
Frans Lanting/Minden Pictures
Page 166
Intestinal tracts adapted from J. G. Fleagle, *Primate Adaption and Evolution*, Academic Press, New York, 1988
Page 168
Adapted from J. G. Fleagle, *Size and Scaling in Primate Biology* (ed. W. L. Jungers), Plenum, New York, 1985

Pages 171 and 172
Adapted from J. Terborgh, *Five New World Primates: A Study in Comparative Ecology*, Princeton University, Princeton, New Jersey, 1983
Page 173
Martyn Colbeck/Charles Munn
Pages 174 and 175
Charles Janson
Page 178
André Bärtschi
Page 180
Adapted from Heinrich Walter, *Vegetation of the Earth and Ecological Systems of the Geo-biosphere*, Springer-Verlag, New York, 1985
Page 182
Adapted from J. Terborgh and C. P. van Schaik, *Organization of Communities Past and Present* (eds. J. H. R. Gee and P. S. Giller), Blackwell, London, 1987
Page 183
Adapted from C. M. Hladik, *Coping with Uncertainty in Food Supply* (eds. I. de Garine and G. A. Harrison), Clarendon, Oxford, 1988
Page 185
Michael and Patricia Fogden
Page 189 top
Michigan Historical Collections, Bently Historical Library, University of Michigan, Ann Arbor
Page 189 bottom
André Bärtschi
Page 191 left
Loren McIntyre
Page 191 right
Brazilian Institute for Space Research (INPE)
Page 192
J. Compton Tucker/NASA
Pages 193 and 194
Loren McIntyre
Page 195
Adapted from *Diversidata 2* (1985)

Page 197
Frans Lanting/Minden Pictures
Page 199
Adapted from D. Western, *Primate Ecology and Conservation* (eds. J. G. Else and P. C. Lee), Cambridge University, Cambridge, 1986
Page 200
T. G. Laman/Anthro Photo
Page 201
Adapted from J. Terborgh and B. Winter, *Conservation Biology: An Evolutionary-Ecological Approach* (eds. M. E. Soule and B. A. Wilcox), Sinauer, Sunderland, Massachusetts, 1980
Page 202
André Bärtschi
Page 204
Bernice Bishop Museum
Page 207
Carl Hansen/Smithsonian Tropical Research Institute
Pages 208 and 210
André Bärtschi
Page 212
Loren McIntyre
Page 214
D. & M. Zimmerman/VIREO
Page 216
Macduff Everton
Page 217
André Bärtschi
Page 218
Carl Jordan
Page 220
Adapted from A. Johns, *Biotropica* 20 (1988)
Page 222
Macduff Everton
Pages 223, 224, 225 and 227
Loren McIntyre
Page 229
Peter Kresan
Page 231
André Bärtschi

Index

Other books in the Scientific American Library Series